Directory of Individuals and Institutions Engaged in South Asian Strategic St

Regional Centre for Strategic Studies

**Directory of Individuals
and Institutions
Engaged in South Asian
Strategic Studies**

Published by:

Regional Centre for Strategic Studies
4-101 BMICH, Bauddhaloka Mawatha
Colombo 7, SRI LANKA.
Tel: (94-1) 688601; Fax: 688602; e-mail: rcss@sri.lanka.net
Website: http://www.lanka.net/rcss

© Regional Centre for Strategic Studies 1998

First Published: July 1998
Printed by : Design Systems, Sri Lanka. Tel: (94-1) 691400

*RCSS is grateful to Ford Foundation for supporting publication of
this Directory and other networking & collaborative projects
undertaken by the Centre.*

ISBN: 955-8051-01-2

CONTENTS

Part I: INDIVIDUALS

Part II: INSTITUTIONS

INTRODUCTION

This directory has been compiled by the Regional Centre for Strategic Studies (RCSS) as a part of its initiatives to promote closer contacts, collaboration and networking of the community of individuals and institutions engaged in South Asian strategic studies. It has been conceived as a quick reference guide on individuals and institutions in and outside the region working on subjects of strategic, security and international concern for South Asia.

In compiling this directory, we proceeded with a multidisciplinary and comprehensive understanding of strategic, security and international studies. Hence, individuals and institutions included here represent a wide and diverse range of professional and academic background and expertise. This is in conformity with the growing concern in and outside South Asia that strategic and security studies cannot be confined to conventional security issues, and should include non-traditional subject-areas, particularly factors that contribute to internal conflict and socio-economic and political instabilities.

The directory is based on a survey conducted by the RCSS in 1996. The survey has generated a wider range of database than what is presented here. The database, which is still evolving, has been helpful in our modest efforts to promote networking and collaboration. The directory is by no means a complete list. We plan to update it on a regular basis. We could not reach many individuals and institutions who should have been included. Over the past two years, through many of our projects we have already been able to bring to our network many individuals and institutions who were not covered by the survey. We would deeply appreciate notification of new entries as well as any inaccuracies and changes that may have taken place since the survey so that we are able to further update, develop and correct our database and the next edition of this directory.

The RCSS is grateful to all individuals and institutions in and outside South Asia who took part in the survey and helped us in conducting it. I am grateful to my colleagues John Gooneratne, Minna Thaheer, Dilkie Perera and Anula Wijesinghe without whose contribution publication of this directory could not be possible. I must express a special word of gratitude and appreciation to Minna Thaheer who joined the Centre at a crucial stage of work for the directory and almost single-handedly and most painstakingly took the burden of bringing it out much faster than it could be otherwise possible. Responsibility for any errors and lapses remains with me.

Iftekharuzzaman
Executive Director

The Regional Centre for Strategic Studies (RCSS) is an independent, non-profit and non-governmental organization for collaborative research, networking and interaction on strategic and international issues pertaining to South Asia. Set up in 1992, the RCSS is based in Colombo, Sri Lanka.

The RCSS is a South Asian forum for studies, training and multi-track dialogue & deliberation on issues of regional interest. All activities of RCSS are designed with a South Asia focus and are usually participated by experts from all South Asian countries. The Centre is envisaged as a forum for advancing the cause of co-operation, security, conflict resolution, confidence building, peace and development in the countries of the South Asian region.

The RCSS serves its South Asian and international constituency by: a) networking programmes that promote interaction, communication and exchange between institutions and individuals within and outside the region engaged in South Asian strategic studies; b) organizing regional workshops & seminars and sponsoring & co-ordinating collaborative research; and c) disseminating output of the research through publications which include books, monographs and a quarterly newsletter. The RCSS facilitates scholars and other professionals of South Asia to address, mutually and collectively, problems and issues of topical interest for all countries of the region.

Queries may be addressed to:

Executive Director
Regional Centre for Strategic Studies
4-101 BMICH, Bauddhaloka Mawatha
Colombo 7, SRI LANKA.
Tel: (94-1) 688601; Fax: 688602; e-mail: edrcss@sri.lanka.net
Website: http://www.lanka.net/rcss

GUIDE TO USERS

Part I: INDIVIDUALS

The names of individuals are listed alphabetically under country and city, also arranged in alphabetical order. In case of individuals from outside South Asia, names are listed alphabetically under country heading only. The format of presentation is as follows:

SURNAME, First Name(s) (Sex/Age) age in January 1996
Academic Position/Profession:
Address:
Tel: country code-area code-number (W for work, H for home); Fax; E-mail
Specialisation:
Areas of Current Research:
Main Work at Hand:
Recent/Forthcoming Publications: (sample titles only)

Part II: INSTITUTIONS

South Asian institutions are listed alphabetically under country and city. Institutions from outside the region are in alphabetical order under country heading only. The information are presented as follows:

Name of Institution (Abbreviations):
Address:
Tel: Fax: E-mail:
Chief Executive:
Other Key Personnel:
Number of Professional Staff:
Areas of Research:
Main Work at Hand:
Publications: Journal(s); Books: Occasional Papers; Recent/Forthcoming Titles; Others.
Seminars and Conferences: Frequency; Participation
Scholarship, Fellowship, Training etc. Title of Programme; Subject(s); Number; Duration; Eligible Participants; Funded by
Library Collections: Books; Periodicals; Newspapers; Others
Year of Establishment:
Status:

LIST OF ABBREVIATIONS

ACDIS	Program in Arms Control, Disarmament & International Security
AP	Andhra Pradesh
ASRC	American Studies Research Centre
BCAS	Bangladesh Centre for Advanced Studies
BELA	Bangladesh Environmental Lawyers' Association
BIDTI	Bandaranaike International Diplomatic Training Institute
BIISS	Bangladesh Institute of International and Strategic Studies
BILIA	Bangladesh Institute of Law and International Affairs
BUP	Bangladesh Unnayan Parishad (Bangladesh Development Council)
CAC	Centre for Analysis and Choice
CAS	Centre for African Studies
CDD	Centre for Democratic Development
CDRB	Centre for Development Research, Bangladesh
CEDA	Centre for Economic Development & Administration
CEPRA	Centre for Policy Research and Analysis
CETS	Centre for Economic & Technical Studies
CISS	The Canadian Institute of Strategic Studies
CNAS	Centre for Nepal & Asian Studies
CPD	Centre for Policy Dialogue
CPR	Centre for Policy Research
CSAS	Centre for South Asian Studies
CSC	Christian Study Centre
CSIS	Centre for Strategic Studies
CSRDS	Centre for Socio-Legal Research & Documentation Service
CUS	Centre for Urban Studies
DEAN	Development Associates Nepal
FRIENDS	Foundation for Research on International Environment National Development Security
HRCP	Human Rights Commission of Pakistan
ICES	International Centre for Ethnic Studies
ICPI	International Centre for Peace Initiatives
ICSSR	Indian Council for Social Science Research
IDCJ	International Development Centre of Japan
IDESS	Institute of Development, Environment & Strategic Studies
IDSA	Institute for Defence Studies & Analyses
IIDS	Institute for Integrated Development Studies
IINS	International Institute for Non-Aligned Studies
IIR	Institute of International Relations
INGO	International Non-Governmental Organisation

IOC	Indian Ocean Centre
IPS	Institute of Policy Studies
ISDA	Institute for the Study of Developing Areas
ISSI	Institute for Strategic Studies, Islamabad
J&K	Jammu and Kashmir
JICA	Japan International Cooperation Agency
JNU	Jawaharlal Nehru University
MP	Madhya Pradesh
MITI	Ministry of International Trade and Industry
NCCS	Nepal Centre for Contemporary Studies
NCWA	Nepal Council of World Affairs
NESAC	Nepal South Asia Centre
NEFAS	Nepal Foundation for Advanced Studies
NGO	Non-Governmental Organisation
NPT	Nuclear Non Proliferation Treaty
NWFP	North West Frontier Province
OIC	Ocean Institute of Canada
PASDA	Pakistan Security & Development Association
PIDE	Pakistan Institute of Development Economics
PIEDAR	Pakistan Institute for Environment- Development Action Research
PIIA	Pakistan Institute of International Affairs
PRC	Peace Research Centre
RCSS	Regional Centre for Strategic Studies
SAARC	South Asian Association for Regional Coorporation
SAGAR	South Asia Group for Action & Reflection
SARS	South Asia Research Society
SDPI	Sustainable Development Policy Institute
SIRSS	School of International Relations & Strategic Studies
TN	Tamil Nadu
UGC	University Grants Commission
UP	Uttar Pradesh
WB	West Bengal

Part I
INDIVIDUALS

CAMILLERI, Joseph A. (M/51)
School of Politics, La Trobe University
Bandoora VIC 3083, Australia.
Tel: 61-03-94792698(W)
Fax: 61-03-94791997(W)

Specialisation : International Relations.

Areas of Current Research:
Regional Security; Multilateral Institutions; Regionalization and Globalization.

Recent/Forthcoming Publications:
The Cold War... and After: A New Period of Upheaval in World Politics
(1995); *The Asia-Pacific in the Post-Hegemonic World, Pacific Cooperation
Building: Economic and Security Regions in the Asia-Pacific Region,* (Boulder, Westview Press 1995); "Human Rights, Cultural Diversity and Conflict
Resolution: The Asia Pacific Context", *Pacifica Review,* (Vol.6, No.2, 1994).

HUDSON-RODD, Nancy (F/46)
Centre for Development Studies, Edith Cowan University
Churchlands Campus, Western Australia 6018, Australia.
Tel: 619-273-8590(W)
Fax: 619-273-8699(W)

Specialisation : Human Geography.

Areas of Current Research:
Sense of Place, Land Rights, Indigenous People, Intra-State Armed Conflict.

Main Work at Hand:
Women in Crisis.

JAYASURIYA, Laksiri (M/64)
Dept. of Social Work & Social Administration
University of W.A., Nedlands 6009WA, Australia.
Fax: 61-9-380-1070(H)

Areas of Current Research:
Comparative Social Policy/Developing Countries; Ethnic Affairs,
Multiculturalism; Sri Lankan Social & Political Development.

Main Work at Hand:
Changing Face of Sri Lankan Politics; The Sri Lankan Welfare State -
Retrospect & Prospect; Australian Multiculturalism & Citizenship.

Recent/Forthcoming Publications:
"Emigration & Settlement in Australia: An Overview & Critique of
Multiculturalism", in N. Caramon (ed.) *Immigration & Integration in Post
Industrial Society* (Macmillan, London).

THAKUR, Ramesh (M/47)
Peace Research Centre, Research School of Pacific & Asian Studies
Australian National University, Canberra ACT 0200, Australia.
Tel: 61-6-249 3098(W)
Fax: 61-6-249 0174(W)
E-mail: thakur@coombs.anu.edu.au

Specialisation : International Relations.

Areas of Current Research:
United Nations Peacekeeping; Asian Pacific Security; Indian Foreign Policy.

Recent/Forthcoming Publications:
The Politics & Economics of India's Foreign Policy (New Delhi & NY: C.
Hurst, Oxford University Press & St Martin's Press, 1994); *The Government
& Politics of India* (London: Macmillan, 1995); *The United Nations at Fifty:
Retrospect & Prospect* (Dunedin University Press, 1996).

BANGLADESH

■ *Chittagong*

AHMED, Tofail (M/41)
Associate Professor, Dept. of Public Administration,
University of Chittagong, Chittagong, Bangladesh.
Tel: 880-31-210131(W) 880-31-650177(H)
Fax: 880-31-210141(W)

Specialisation : Political Economy.

Areas of Current Research:
Social Development Policy & Planning; Local Government & Rural Development; Civil Society & Democracy.

Main Work at Hand:
Decentralization & the Local State Under Peripheral Capitalism; Decentralized District Planning.

Recent/Forthcoming Publications:
"Gandhi's Approach to Rural Development (1995)", *Journal of Asiatic Society of
Bangladesh*; "Community Participation: & Literacy", *Journal of Rural Development*; Rural Development & Comilla Approach: Option for Bangladesh.

ALAM, M. Fashiul (M/43)
Professor, Dept. of Management
University of Chittagong, Bangladesh.
Tel: 880-31-210131 Ext.4304(W)

Specialisation : Industrial Relations.

Areas of Current Research:
Industrial Relations; Human Resource Management; Collective Bargaining;
Trade Unionism.

Main Work at Hand:
Env. Forces Affecting Trade Unionism in Bangladesh; Trends in International
Relations in South Asia.

Recent/Forthcoming Publications:
Profile of Bangladesh Community in the UK; Management Structure &
Characteristics of the Enterprises of the Self-Employed Bangladeshis in UK

BARUA, Benli Prasad (M/59)
Department of Political Science, University of Chittagong
Chittagong, Bangladesh.
Tel: 880-31-210141-44(W)

Specialisation : Political Science.

Areas of Current Research:
Ethnic Politics; Oriental Political Thought; Bangladesh Politics.

Main Work at Hand:
Ethnicity & the Integrational Crisis: The Case of Chittagong Hill Tracts,
Bangladesh.

Recent/Forthcoming Publications:
The Roots of Crisis in the Chittagong Hill Tracts, Bangladesh; Tagore, India
& Bangladesh; Buddhist Festivals & Rituals in Bangladesh.

BEGUM, Syeda Tahera (F/49)
Department of Economics, University of Chittagong
Chittagong, Bangladesh.
Tel: 880-31-210133-35(W)
Fax: 880-31-210141(W)

Specialisation : Economics.

Areas of Current Research:
Rural Women & Economic Development.

Main Work at Hand:
Role of Women in Marine Fishing.

Recent/Forthcoming Publications:
Female Participation: in Agricultural Activities & Fertility (U.G.C., Dhaka);
Economic Evaluation of Shrimp Cultivation (Ford Foundation, Dhaka);
Women in Marine Fishing (FAO, Mumbai).

BEGUM, Lutfun Nahar (F/27)
Lecturer, Dept. of Sociology
University of Chittagong, Chittagong, Bangladesh.
Tel: 880-31-619286(H)

Specialisation : Sociology.

Areas of Current Research:
Women in Development; Ethnic Studies; Tribal Society.

Main Work at Hand:
Impact of National Politics on Student Politics in the Context of Bangladesh;
Women's Participation in Alleviating Poverty of Bangladesh.

Recent/Forthcoming Publications:
"Women in Politics in the Context of South Asia".

BHUIAN, Monoar Kabir (M/35)
Associate Professor, Dept. of Political Science
University of Chittagong, Chittagong, Bangladesh.
Tel: 880-31-210141-44(W)

Specialisation : Political Science.

Areas of Current Research:
International Relations of East Asia; Political Economy of the Developing
World; Civil-Military Relations & Democratization.

Main Work at Hand:
Bangladesh: Politics of Military Rule & Religion; Explaining Democratic
Stability in India.

Recent/Forthcoming Publications:
"Politico-Economic Limitations & Fall of the Authoritarian Government in
Bangladesh", *Armed Forces & Society*, (Summer 1995).

CHOUDHURY, Ahmed Fazle Hasan (M/51)
Professor, Dept. of Sociology
Chittagong University, Chittagong, Bangladesh.
Tel: 880-31-714932

Specialisation : Anthropology.

Areas of Current Research:
Ecological Issues; Women in Development.

Main Work at Hand:
Health Problems Among Coastal Women in Bangladesh; Socio Legal Status of Tribal Women of Bangladesh.

Recent/Forthcoming Publications:
Environmental Depletion in the CHT Region.

CHOWDHURY, H. Mahfuzul (M/42)
Associate Professor, Dept. of Political Science
Chittagong University, Chittagong, Bangladesh.
Tel: 880-31-210141-44(W)

Specialisation : Law & Public Policy.

Areas of Current Research:
US Foreign Policy; Political Development; Ethnicity & National Security.

Main Work at Hand:
Aid & Security in US Third World Policy; Ethnic Conflict & National Integration in India, Bangladesh & Sri Lanka.

Recent/Forthcoming Publications:
"Ethnic Conflicts & National Integration in India", *Regional Studies*, Spring, 1994, Vol. XII, 2; "Aid and Security in US Policy Towards Developing Nations", *Pakistan Journal of American Studies*, Vol. 13, No. 1, 1995.

CHOWDHURY, Abdul Mannan (M/45)
Professor, Dept. of Economics, University of Chittagong
Chittagong, Bangladesh.
Tel: 880-31-210133(W)

Specialisation : Economics.

Areas of Current Research:
Rural Landlessness, Food Policy; Foreign Investment, MNC's; Unemployment, Trade Unionism.

Main Work at Hand:
A Critical Appraisal of the Uruguay Round GATT in the Context of Developing Countries; Underutilization of Labour in Bangladesh: An Application of the Houser Concept.

Recent/Forthcoming Publications:
"The Impact of Irrigation on Agrarian Labour Relations", *Chittagong University Studies*, Vol.XV, No.1; *Relative Performances of Nationalised Commercial Banks, 1971-1989: A Critical Analysis* (Bankers, Institute of Bankers, Bangladesh).

HOSSAIN, Mohammad Abdul (M/33)
Assistant Professor, Dept. of Economics
University of Chittagong, Bangladesh.
Tel: 880-31-210133-35(W)

Specialisation : Economics.

Areas of Current Research:
International Trade; Macroeconomics; Econometrics.

Main Work at Hand:
Price & Income Elasticities of Imports & Exports of the Developing Countries: The Case of Bangladesh

Recent/Forthcoming Publications:
Effects of Devaluation on Trade Balance: Empirical Evidence; A Critique of Classical Macroeconomic Models; Changes in the Composition of Imports & Exports of Bangladesh.

HUQUE, Mahmudul (M/46)
Dept. of History, University of Chittagong
Chittagong, Bangladesh.
Tel: 880-31-714923(W)

Specialisation : History.

Areas of Current Research:
US-South Asia Relations; Regional Conflicts in South Asia; Nuclear Proliferation in South Asia.

Main Work at Hand:
Trends in Regional Relations Among South Asian Countries; US Foreign Aid Policy After the Cold War.

Recent/Forthcoming Publications:
"US Indian Relations During the Kennedy Administration: Unfulfilled Expectations", *Pakistan Journal of American Studies,* 11:2 fall 1993; "US Relations with India & Pakistan After the Cold War" (1995 Working Paper no. 1, CISSM, School of Public Affairs, Univ. of Maryland), A Review of US Policy During the India-Pakistan War of 1965.

HUSSAIN, Hayat (M/49)
Dept. of History, University of Chittagong
Chittagong, Bangladesh.
Tel: 880-31-714923(W) 31-613151(H)
Fax: 880-31-210141(W)

Specialisation : History.

Areas of Current Research:
Bengal in the Eighteenth Century; Tribal Studies; Middle East in Modern Times.

Main Work at Hand:
Political & Constitutional Rights of the Hillmen of CHT.

Recent/Forthcoming Publications:
Bangladesh & the Crisis of Democracy (in Bengali) 1985; *Surya Sen & the Chittagong Revolt* (in Bengali) 1985.

IMAM, A.F. Ali (M/43)
Professor & Chairman, Dept. of Sociology
Chittagong University, Chittagong, Bangladesh.
Tel: 880-31-714923(W)
Fax: 880-31-210141(W)

Specialisation : Sociology.

Areas of Current Research:
Social Stratification; Social Inequalities & Integration;
Educational & Occupational Inequalities.

Main Work at Hand:
Social Structure of a Christian Community.

Recent/Forthcoming Publications:
Social Sratification Among the Muslim and Hindu Community (1992); *Changing Social Stratification in Rural Bangladesh* (1993); "Social Stratification, Interaction & Integration Among the Muslims & the Hindus in a Bangladesh Village" in Grover, V. (ed) *Government & Politics in South Asian Countries,* 1996.

ISLAM, Muinul (M/45)
Professor, Department of Economics
University of Chittagong, Chittagong, Bangladesh.
Tel: 880-31-210133(W)

Specialisation : Economics.

Areas of Current Research:
Economic Development; Migration.

Main Work at Hand:
The Trapped Sojourners: A Study of Bangladeshi Immigrants in Britain;
Political Economy of Corruption.

Recent/Forthcoming Publications:
"Employment Generation in Chittagong City", *UNCHS,* March 1995.

KHALED, A.M.M. Saifuddin (M/43)
Department of History, University of Chittagong
Chittagong, Bangladesh.
Tel: 880-31-714923-4367(W)

Specialisation : American History & Diplomacy.

Areas of Current Research:
US Diplomacy in South Asia; US Cold War Strategies in Europe, South Asia
& the Middle East Following World War II; American Policy Regarding
India-Pakistan Disputes (Kashmir, Bangladesh).

Main Work at Hand:
American Policy in South Asia During the Bangladesh War of Independence,
1971; American Arms for Pakistan, 1954-1971.

Recent/Forthcoming Publications:

"US Role in Early Stages of Kashmir Conflict", *Regional Studies* 12, No.1
(Islamabad, Pakistan, Winter 1993-94); "Geopolitics & Regional Reality:
The US & the Kashmir Dispute", *Journal of the Asiatic Society of Bangladesh*, June 1994.

KHANAM, Johora (F/45)
Associate Professor, Dept. of Political Science
University of Chittagong, Chittagong, Bangladesh.
Tel: 880-31-21014-44(W)

Specialisation : Political Development in Bangladesh.

Areas of Current Research:
Bangladesh Politics; Rural Development.

Main Work at Hand:
The Formative Phase of Bangladesh & Development of the Opposition
Parties: An Analysis; From Autonomy to Independence: An Analytical Study
of the Development of the Awami League.

Recent/Forthcoming Publications:
"Identity Crisis & the Problem of National Integration in Bangladesh During
the Awami League Rule: A Study of the Ethnic Minority Groups in
Chittagong Hill Tracts", *Bangladesh Political Studies,* Vol. 15, 1993; "The
Leftists in Bangladesh Politics: Crisis & Sequences", *Asian Profile*, 235, Oct
95; Parliamentary Democracy in Bangladesh & the Practice 1972-75.

MAHFUZ PARVEZ, A.K.M. Mahfuzul Haque (M/29)
Assistant. Professor, Dept. of Political Science, University of Chittagong
Chittagong, Bangladesh.
Tel: 880-031-503173(W)

Specialisation : Government & Politics.

Areas of Current Research:
Peace & Conflict Studies in South Asian Perspective; Good Governance & Democracy in South Asia; Issues of Regional Co-operation.

Main Work at Hand:
Roots of Democracy in South Asian States; Parliamentary Democracy in Post-Liberation Bangladesh.

Recent/Forthcoming Publications:
Bangladesh Politics: Post-Liberation Era (Edited Book, AP. Dhaka); "Military Intervention in Politics: Bangladesh Perspective", *Political Science Studies*; Feminism in Third World.

MUHAMMAD, A. Hakim (M/46)
Professor, Dept. of Political Science
University of Chittagong, Chittagong, Bangladesh.
Tel: 880-31-210141-44(W)

Specialisation : Public Administration.

Areas of Current Research:
Comparative Politics; Bangladesh Politics; Bangladesh Administration.

Main Work at Hand:
Legitimacy Crisis; Islam; State in Bangladesh.

Recent/Forthcoming Publications:
"The Mirpur Parliamentary By-Election in Bangladesh", *Asian Survey,* Vol. 34, No.8; "Determinants of Administration in Developing Countries: A Review & a Basis for Analysis"; "The Bangladesh Experience", *Philipine Journal of Public Administration,* Vol. 36, No. 2.

MUSHRAFI, Mokhdum-E-Mulk (M/46)
Professor & Chairman, Dept. of Politics
University of Chittagong, Chittagong, Bangladesh.
Tel: 880-31-210141-44

Specialisation : Political Science.

Areas of Current Research:
Political Culture; Party Process; Nature of Civilizational Evolution & Politics.

Main Work at Hand:
Construing Authoritarianism in Primordial Politics.

Recent/Forthcoming Publications:
Pakistan & Bangladesh: Political Culture and Political Parties.

NIZAMI, Helal Uddin (M/29)
Asst. Professor, Dept. of Accounting
University of Chittagong, Bangladesh.
Tel: 880-31-210131 Ext.328(W), 619286(H)

Specialisation : Finance.

Areas of Current Research:
International Finance Policy; Capital Market Development of South Asian
Countries; Impact of Trade Agreements in South Asian Business in the Light
of GATT & Uruguay Round.

Main Work at Hand:
Profitability Performance of Foreign Banks Working in Bangladesh

Recent/Forthcoming Publications:
Role of Foreign Banks in Financing Industrial Projects: A Study on ADB &
IMF; Role of Multinational Companies in the Economic Development of
Bangladesh.

OSMAN, Ferdous Arfina (F/27)
Lecturer, Dept. of Public Administration
University of Chittagong, Chittagong, Bangladesh.

Specialisation : Public Administration.

Areas of Current Research:
Public Policy; Public Sector Productivity; Human Resource Development.

Main Work at Hand:
Dynamics of Public Policy Making Process: A Case Study of Health Policy
of Bangladesh; Privatization Experience of Bangladesh Since 1975.

Recent/Forthcoming Publications:
"Nature of Governance & Structural Adjustment Policy Reform in
Bangladesh", *Chittagong University Studies, Social Science* Vol. XV.

RAHMAN, Mohammad Mafizur (M/33)
Dept. of Economics, University of Chittagong
Chittagong, Bangladesh.
Tel: 880-31-210133-35(W)

Specialisation : Economics.

Areas of Current Research:
Development Economics; International Economics; Macro Economics.

Main Work at Hand:
Human Resources Development Situation in Bangladesh.

Recent/Forthcoming Publications:
Do Transnational Corporations Benefit or Harm the Economic Development
Situation in Bangladesh?

ULLAH, M. Mahbub (M/46)
Professor, Dept. of Sociology
University of Chittagong, Chittagong, Bangladesh.
Tel: 880-31-210131(W) 615506(H)

Specialisation : Sociology.

Areas of Current Research:
Social Structure & Social Change; Society & Politics.

Main Work at Hand:
The Growth of Nationalism & the Emergence of Bangladesh as a Nation
State.

Recent/Forthcoming Publications:
The History of the Emergence of Bangladesh as a Nation-State.

■ Dhaka

AFROZE, Shaheen (F/36)
Senior Research Fellow
BIISS, 1/46, Elephant Road
Dhaka 1000, Bangladesh.
Tel: 880-2-406234(W) 603022(H)
Fax: 880-2-832625(W)
E-mail: biiss@drik.bgd.toolnet.org

Specialisation : International Relations.

Areas of Current Research:
Foreign Policy & Security of Small States; South Asian Studies;
Inter-State Relations in South Asia.

Main Work at Hand:
Managing the Traumas of Democratic Transition: The Case of South Asia.

Recent/Forthcoming Publications:
"Nuclear Rivalry & Non-Nuclear Weapon States in South Asia: Policy
Contingency Framework", *BIISS Journal,* Vol.16, No.4,1995; Nepal: Domes-
tic Politics & Foreign Relations.

AHMAD, Muzaffer (M/58)
Institute of Buisness Administration, Dhaka University
Dhaka 1000, Bangladesh.

Tel: 880-2-505551(W), 812022(H)
Fax: 880-2-811138(H)

Specialisation : Economics.

Areas of Current Research:
Human Development; Structural Reform; Political Economy.

Main Work at Hand:
Health of Our Nation; Essays in Political Economy.

Recent/Forthcoming Publications:
Political Economy of Bank Loan Default; Quality Status in Health & Education; Poverty Attenuation in the New World Order.

AHMAD, Nasreen (F/46)
Dept of Geography, University of Dhaka, Science Annexe Building, Ramna, Dhaka 1000, Bangladesh.
Tel: 880-2-327037, 815572(H)

Specialisation : Geography; Demography.

Areas of Current Research:
Population/Women (Gender); Environment.

Main Work at Hand:
Determinants of Landlessness in Bangladesh: A Geographical Analysis.

Recent/Forthcoming Publications:
Problems Faced by Women-Headed Households Residing in Embankment Settlements, Nov.1995; Prostitutes & their Environment in Bangladesh: A Geographical Perspective.

AHMAD, Qazi Kholiquzzaman (M/52)
Chairman, BUP, 33, Road 4,Dhanmondi R.A.
P.O. Box 5007 (New Market), Dhaka, Bangladesh.
Tel: 880-2-508097(W)
Fax: 880-2-867021(W)
E-mail: qka.bup@drik.bgd.toolnet.org

Specialisation : Economics.

Areas of Current Research:
Resource Management; Employment Generation & Poverty Alleviation; Environment.

Main Work at Hand:
Regional Study on Water Based Integrated Development; Socio-Economic Implications of Agricultural Mechanization in Bangladesh.

Recent/Forthcoming Publications:
The Implications of Climate & Sea Level Change for Bangladesh, co-ed,
(Kluwer Academic Publishers, The Netherlands, 1996); *Democracy &
Development: Must there be Winners & Loosers?*, (BUP, Dhaka 1995);
*Converting Water into Wealth: Regional Co-operation in Harnessing the
Eastern Himalayan Rivers,* (BUP/Academic Publishers, Dhaka, 1994).

AHMED, Abu Taher Salahuddin (M/34)
BIISS, 1/46, Elephant Road
Dhaka 1000, Bangladesh.
Tel: 880-2-406234(W)
Fax: 880-2-832625(W)
E-mail: biiss@drik.bgd.toolnet.org

Specialisation : International Relations.

Areas of Current Research:
Security & Strategic Affairs; Inter-State Relations; Conflict; Environmental
Studies.

Main Work at Hand:
"Use of Energy in South Asia: Problems & Prospects"; Nuclear Proliferation
& Deterrence in South Asia: The Role of the USA; Maritime Boundary
Problems Between Bangladesh & India.

Recent/Forthcoming Publications:
"Myanmar: Road to Democracy or East Asian Model?" *BIISS Journal*;
Indian Naval Expansion: A Conspectus, Nov. 1994.

AHMED, Emajuddin (M/62)
Professor, Dept. of Political Science, University of Dhaka
Dhaka 1000, Bangladesh.
Tel: 880-2-868383(W), 501110(H)
Fax: 880-2-865583(W)

Specialisation : Political Science.

Areas of Current Research:
Role of Civil Bureaucracy in South Asia; The Military & Democracy in the
Third World; Foreign Policy & International Relations.

Main Work at Hand:
Problems & Prospects of Democratic Order in Bangladesh; Problematics of
Local Government.

Recent/Forthcoming Publications:
Future of Democracy (Dhaka: Bangla Academy, 1988); *Military Rule & Myth
of Democracy* (Dhaka: University Press Ltd.); *Society & Politics In Bangladesh* (Dhaka: Academic Publishers,1989).

AHMED, Fakhruddin (M/64)
Former Foreign Secretary, House No.22, Road No. 7
Dhanmondi R/A, Dhaka 1205, Bangladesh.
Tel: 880-2-9111718(W) 326533(H)
Fax: 880-2-326303(W)

Specialisation : International Relations.

Areas of Current Research:
International & Regional Affairs.

Main Work at Hand:
Election Observation.

AHMED, Imtiaz (M/37)
Associate Professor, Dept. of International Relations
University of Dhaka, Dhaka 1000, Bangladesh.
Tel: 880-2-505769(W) 417108(H)
Fax: 880-2-833355(H)
E-mail: imtiaz@bangla.net

Specialisation : Political Science.

Areas of Current Research:
Theories of International Relations; South Asian Politics; Refugees &
Migrations.

Main Work at Hand:
The Plight of the Environmental Refugees; The Efficiency of the Nation-
State in South Asia.

Recent/Forthcoming Publications:
State & Foreign Policy: India's Role in South Asia, 1993; Refugees &
Security: The Experience of Bangladesh, 1996; On Feminist Methodology,
1995.

AHMED, Israt (F/28)
Dept. of Anthropology, Jahangirnagar University
Savar, Dhaka, Bangladesh.
Tel: 880-2-381321(H)

Specialisation : Anthropology.

Areas of Current Research:
Environment; Women in Development; Health & Nutrition.

Main Work at Hand:
Book on Political Anthropology; Research on Women's Own Perception of
Womenhood in Bangladesh.

Recent/Forthcoming Publications:
"Influence of Women's Development Programmes on Women's Status &
Fertility in Bangladesh: Evidence from Focus Group Discussions"
with M. Kabir, *Eastern Anthropologist,* Vol.47, No.4, Oct-Dec 1994;
"Women in Development: Experience of NGO's, In the Kapasia Thana of
Gazipur District" with S.M. Nurul Alam & Rasheda Akhtar, *Empowerment,
Journal of Women for Women,* Vol.2, 1996, Dhaka; "Anthropological Contri-
bution to the Study of Religion: A Review of Selected Issues", *Journal of
Social Science,* Jahangirnagar University, Savar, Dhaka, Vol.15, 1996.

AHMED, Rehnuma (F/38)
Assistant Professor, Dept. of Anthropology
Jahangirnagar University, Dhaka, Bangladesh.

Areas of Current Research:
Gender, Class, Differentiation; Colonialism.

Main Work at Hand:
Establishing Families: Gender, Class & Kinship in Dhaka City.

AHMED, Sabbir (M/29)
Research Associate, BIISS, 1/46, Elephant Road
Dhaka 1000, Bangladesh.
Fax: 880-2-832625(W)
E-mail: biiss@drik.bgd.toolnet.org

Specialisation : Political Science.

Areas of Current Research:
Political Development; Conflict Studies; Political Economy.

Main Work at Hand:
Military Withdrawal from Bangladesh Politics: Continuity or Change?;
Shifts in the Trends of Bangladesh Politics: A Critical Analysis.

Recent/Forthcoming Publications:
"SAARC Preferential Trading Arrangements: A Preliminary Analysis",
BIISS Journal, Vol.16, No.2, 1995; The Mohajir Quaomi Movement in
Pakistan: An Analysis of an Intra-State Conflict.

AHSAN, Rosie Majid (F/53)
Department of Geography, University of Dhaka
Dhaka 1000, Bangladesh.

Specialisation : Geography.

Areas of Current Research:
Urban-Poor; Waste Management; Women's Activity, Migration Development; Environment-Urban & Rural.

Main Work at Hand:
Peri Urban Areas of Bangladesh Towns; Infrastructure Facilities in Dhaka City.

Recent/Forthcoming Publications:
The Invisible Resource: Women & Work in Rural Bangladesh (co-Authored), (West View Press, Boulder, 1987); "Migration Pattern & Process of Female Construction Labour in Dhaka City", *Population Geography* (Dundee, UK).

AKHTAR, Rasheda (F/31)
Dept. of Anthropology, Jahangirnagar University
Savar, Dhaka, Bangladesh.
Tel: 880-2-869414(H)

Specialisation : Sociology.

Areas of Current Research:
Environment; Gender Issues; Health Issues.

Main Work at Hand:
Women's Perception of Womenhood in Bangladesh.

Recent/Forthcoming Publications:
"Women in Development: Experience of NGO's in the Kapasia Thana of Gazipur District", with S.M. Nurul Alam & Israt Ahmed, *Empowerment, A Journal of Women for Women,* Vol. 2, 1996, Dhaka; Women & Children in Disaster.

AKTER BANU, U.A.B. Razia (F/46)
Professor, Dept. of Political Science, University of Dhaka
Dhaka 1000, Bangladesh.
Tel: 880-2-503083(W) 506311(H)
Fax: 880-2-865583(W)

Specialisation : Political Science.

Areas of Current Research:
Religion, Culture & Politics; Peace & Ethnicity; Women & Political Development.

Main Work at Hand:
American Scholars on Islam; Political Culture of Japan & Bangladesh - A Comparative Study.

Recent/Forthcoming Publications:
Islam in Bangladesh (E.J. Brill, 1992); Rights of Women in Bangladesh in Constitution & Religion; Influence of Culture & Religion on Politics.

AMIN, Muhammad Ruhul (M/27)
Lecturer, International Relations
University of Dhaka, Dhaka 1000, Bangladesh.
Tel: 880-2-505769(W)

Specialisation : International Relations.

Areas of Current Research:
South Asian Studies; Religious Fundamentalism; Bangladesh Foreign Policy;
Islam & the West.

Main Work at Hand:
Comparative Study of Religious Fundamentalism in South Asia; Economic
Diplomacy of Bangladesh.

Recent/Forthcoming Publications:
"SAARC: Opportunities & Challenges", *Social Science Review,* Dhaka
University; "Post-Cold War Foreign Policy of Bangladesh: Implications for
US-Bangladesh Relations", *Dhaka University Journal*; "Economic Diplo-
macy of Bangladesh: A Post-Cold War Perspective", *Journal of International
Relations,* Dhaka University.

AMINUL, Haque (M/31)
Centre for Policy Dialogue, 6A Eskaton Garden,
Ramna, Dhaka 1000, Bangladesh.
Tel: 880-2-837055(W)
Fax: 880-2-835701(W)

Specialisation : History & Political Economy.

Areas of Current Research:
Aid Policy & Aid Relations; Governance & Development.

Main Work at Hand:
Governance & Development: The Nature & the Quality of Governance in
Malaysia.

Recent/Forthcoming Publications:
*Small Versus Large Donors: Swedish Aid to Bangladesh 1962-1993, A
Comparative Perspective* (Goteburg University Press, Sweden, 1994).

BEPARI, Nurul Amin (M/38)
Professor, Dept. of Political Science
University of Dhaka, Dhaka 1000, Bangladesh.
Tel: 880-2-505769(W)

Specialisation : Political Science.

Areas of Current Research:
South Asian Studies; Politics of Developing Areas; Bangladesh Politics.

Main Work at Hand:
National Integration Crisis in Bangladesh: The Case of the Chittagong Hill Tracts.

Recent/Forthcoming Publications:
"The Pro-Chinese Communist Movement in Bangladesh", *Journal of Contemporay Asia*; "Maoist Politics in India" *Regional Studies*; "Twenty Years of Bangladesh Politics" *Regional Studies*.

BHUYAN, Ayubur Rahman (M/56)
Dept. of Economics, University of Dhaka
Dhaka 1000, Bangladesh.
Tel: 880-2-505711(W) 500260(H)
Fax: 880-2-863905(W)

Specialisation : Economics.

Areas of Current Research:
International Economics; Public Finance; Monetary Economics.

Main Work at Hand:
Fiscal Reforms in Bangladesh; Effective Rates of Protection in Bangladesh Textile Sector.

Recent/Forthcoming Publications:
A South Asian Customs Union: Prospects & Problems, (Dhaka: E.R.D. Govt of Bangladesh); *Scarcity Premium of Imports & Shadow Price in Bangladesh*, (Dhaka: Planning Com. Govt. of Bangladesh 1985); *Trade, Regions & Industrial Growth: A Case Study of Bangladesh*, (San Francisco, Cal: I.C.E.G. 1993).

CHOUDHURY, R. Abrar (M)
Chairman, Department of International Relations
& Coordinator
Refugee and Migratory Movement Research Unit
Dhaka University
Dhaka 1000, BANGLADESH.
Tel: 880-2-9661900-59(W); 503519 (H)
Fax: 880-2-817962;
E-mail: cabrar@bangla.net

Specialisation: International Relations

Areas of Current Research:
Asian Studies; Refugee and Migratory Movements; Environment and Security; Civil-Military Relations

Main Work at Hand:
Refugee Law & Legislation in Bangladesh; Plight of Internally Displaced
Persons in Bangladesh; Environment and Law.

Recent/forthcoming Publications:
Articles on environment and the military; "Voluntary Repatriation of
Rohingya Refugees"; "International Agreements and Environmental Manage-
ment Follow-up in Bangladesh"; "Issues and the Constraints in the Repatria-
tion/Rehabilitation of the Rohingya and Chakma Refugees and the Biharis";
"Defence Expenditures in Bangladesh: Prospects for Reduction".

CHOUDHURY, Dilara (F/50)
Dept. of Government & Politics, Jahangirnagar University
Savar, Dhaka 1342, Bangladesh.
Tel: 880-2-9330185(W) 819499(H)

Specialisation : Constitutional Development.

Areas of Current Research:
South Asian International Systems; Constitutionalism & Democracy; Women
in Development.

Main Work at Hand:
The US & Conflict Resolution in South Asia; Women in Development
in Bangladesh.

Recent/Forthcoming Publications:
"The Evolution of Judiciary in Bangladesh", *Asian Studies,* Vol.15, 1996;
The Foreign Policy of Bangladesh; The Post-European World Order &
Conflict Resolution in South Asia (Area Study Centre for Europe, University
of Karachi).

D'COSTA, Dorothy (F/26)
Centre for Policy Dialogue, 6A Eskaton Garden
Dhaka 1000, Bangladesh.
Tel: 880-2-837055(W) 880-2-889250(H)
Fax: 880-2-835701(W)

Specialisation : International Relations.

Areas of Current Research:
Human Rights & Development; Governance.

DAS, Subash Chandra (M/45)
Professor, Dept. of Geography, Jahangirnagar University
Savar, Dhaka, Bangladesh.

Specialisation : Oceanography.

Areas of Current Research:
Law of the Sea & the Conflict of Exploitation of Sea Resources;
Environmental Impact of Development Activities in the Coastal; Appraisal of
Coastal Resources in Bangladesh.

Main Work at Hand:
Environmental & Coastal Resource Management in Bangladesh; Disaster
Management in Bangladesh.

Recent/Forthcoming Publications:
Geography of Coastal Environment in Bangladesh (Jahangirnagar University,
Dhaka, 1996); Coastal Land Use Change & the Resulting Impact on Environment.

EUSUF, Ammatuz Zohra (F/48)
Associate Professor, Dept. of Geography, University of Dhaka
Dhaka 1000, Bangladesh.
Tel: 880-2-505740(W) 832043(H)

Specialisation : Geography.

Areas of Current Research:
Cultural Geography; Economic Geography.

Main Work at Hand:
"Tribal & Non-Tribal Population of Chittagong Hill Tracts".

Recent/Forthcoming Publications:
"Variations of Quality of Life of Tribal People in Hill Tracts Region of
Bangladesh: A Spatial Analysis", *Journal of Rural Development,* Vol.26,
No.1, Jan 1966, pp.49-72, B.A.R.D., Camilla, Bangladesh; "Female Educa-
tion Scenario in Bangladesh: A Spatio Temporal Perspective 1961-1991",
Dhaka University Journal of Science, Vol 43, No 1, Jan 1995.

HAFIZ, M. Abdul (M/59)
106 New DOHs , Block 'A', Mohakhali
Dhaka 1206, Bangladesh

Specialisation : Politics & International Relations

Areas of Current Research:
Environmental Security; Security of Developing Countries; Peace Research
& Conflict Studies.

Main Work at Hand:
Environmental Degradation & Intra/Inter State Conflict in Lower Ganges
Basin; Group Interests in Bangladesh Military.

Recent/Forthcoming Publications:
"The Challenges to Security Studies in a Phase of New Orientation",

Perspective, Verlag Ruegger AG, Zurich; "Rethinking South Asia's Security" *Mirpur Papers,* Issue No.3, 1995; "Environmental Degradation & Intra/Inter Conflicts in Bangladesh", *Occasional Paper* No. 6, Environment & Conflict Project, Swiss Peace Foundation, 1993.

HALIM, Abdul (M/43)
Professor, Dept. of International Relations
University of Dhaka, Dhaka 1000, Bangladesh.
Tel: 880-2-505769(W)

Specialisation : International Relations.

Areas of Current Research:
International Relations Theory; Research Methodology; Foreign Policy of Bangladesh.

Main Work at Hand:
Strategic & Goepolitical Importance of Bangladesh; An Inquiry into the Causes of the Fall of Communism

Recent/Forthcoming Publications:
Is the UN an Instrument of US National Policy?; The Paradigmatic Status of International Relations; "Foreign Affairs: Safeguarding National Interest", in *Zia Episode in Bangladesh Politics*, (1996 with Kamal Uddin Ahmed).

HAQUE, Azizul (M)
Professor of Government & Politics, Jahangirnagar University
Savar, Dhaka 1342, Bangladesh.

Specialisation : Politics.

Areas of Current Research:
South Asia; External Relations; National Insurgencies.

Main Work at Hand:
Trends in Pakistan's External Policy; Bangladesh: Society, Politics, Democracy.

Recent/Forthcoming Publications:
"Zia Politics & Strategies: A Peep into their Limitations", *Asian Studies,* June 1994.

HAQUE, Ehsanul (M/30)
Assistant Professor, Dept. of International Relations
University of Dhaka, Dhaka 1000, Bangladesh.
Tel: 880-2-505769(W)
Fax: 880-2-865583(W)

Specialisation : International Relations.

Areas of Current Research:
International Security; South East & North East Asian Affairs; South Asian Affairs.

Main Work at Hand:
The Kashmir Factor in Indo-Pakistan Relations; Nuclear Nightmare in the Korean Peninsula.

Recent/Forthcoming Publications:
"Bangladesh & the United States: Dimensions of an Evolving Relationship", in Abdul Kalam (ed.), *Bangladesh: Internal Dynamics & External Linkages* (Dhaka: University Press Ltd., 1996).

HAQUE, Mahfuzul (M/43)
Senior Assistant Secretary
Ministry of Environment & Forests, Room 1303
Building 6, Bangladesh Secretariat
Dhaka, Bangladesh.
Tel: 880-2-860551(H)

Specialisation : International Relations.

Areas of Current Research:
Ethnic Groups & Anthropological Studies; Mountain Environment & Life of the People Therein.

Main Work at Hand:
Ethnic Insurgency & National Integration; A Tale of Refugees: Rohingyas in Bangladesh.

Recent/Forthcoming Publications:
Global Warming & Sea Level Rise: Bangladesh Case; Street Children of Bangladesh.

HASANUZZAMAN, Al Masud (M/40)
Dept. of Government & Politics, Jahangirnagar University
Savar, Dhaka, Bangladesh.

Specialisation : Political Science.

Areas of Current Research:
Bangladesh Politics & Development; Legislature & Legislative Politics.

Main Work at Hand:
Role of Opposition in Bangladesh Politics.

Recent/Forthcoming Publications:
Institution Building Process in Bangladesh 1971-96; "Legislative Role of Opposition in Bangladesh & Parliament", *Asian Studies*, Vol.15. 1996; Foreign Policy of Bangladesh.

HASSAN, Shahed (M/46)
Professor & Chairman, Dept. of Anthropology,
University of Dhaka
Dhaka 1000, Bangladesh.
Tel: 880-2-411162(H)
Fax: 880-2-865583(W)

Specialisation : Anthropology.

Areas of Current Research:
Ethnic Issues; Ecological Concerns; Indigenous Knowledge System.

Main Work at Hand:
Rehabilitation of the Chakma Refugees of Chittagong Hill Tracts.

HASSAN, Shaukat (M/43)
Head, Dept. of Environmental Studies, North South University
12 Kemal Ataturk Avenue, Banani, Dhaka 1213, Bangladesh.
Tel: 880-2-885369(H)

Specialisation : International Relations.

Areas of Current Research:
Strategic Concerns; Environmental Issues; Peace Research.

Main Work at Hand:
Environmental Security; State Capacity & Civil Violence: The Case of Bihar.

Recent/Forthcoming Publications:
Above Title to be Published Jointly by the American Academy of Arts &
Science, Boston, and the Centre for Conflict Studies, Toronto as Occasional
Papers.

HOSSAIN, Delwar (M/28)
Lecturer, Dept. of International Relations
University of Dhaka, Dhaka 1000, Bangladesh.

Specialisation : International Relations.

Areas of Current Research:
Security Studies; State & Civil Society (Particulary Environment, Migration,
Economic Issues); Diplomacy, Regional Co-operation.

Main Work at Hand:
Democracies & Interstate Conflict in South Asia; State & Nation in the Post-
Cold War Era.

Recent/Forthcoming Publications:
"US-Japan Trade Friction in the Post Cold War Era: Issues & Ramifications"
BIISS Journal, Vol 16, # 3, 1995; "Palestinians-Israeli Conflict After the Oslo

Process: Changes & Continuity", *Journal of International Relations.* Vol 2, # 2, 1995; "Combating Terrorism Without Violating Human Rights: Some Options Revisited", *BIISS Journal, V*ol 15,# 3, 1994.

HOSSAIN, Golam (M/41)
Dept. of Government & Politics, Jahangirnagar University
Savar, Dhaka 1342, Bangladesh.

Specialisation : Political Science.

Areas of Current Research:
Civil Military Relations in Bangladesh; Democratization in Bangladesh.

Main Work at Hand:
Comparative Study of the Performance of Civil & Military Regimes in Bangladesh.

Recent/Forthcoming Publications:
"Bangladesh in 1994: Democracy at Risk", *Asian Survey,* California; "Bangladesh in 1995: Politics of Intransigence", *Asian Survey,* California.

HAQ, Ramjul (M/46)
Dept. of International Relations, University of Dhaka
Dhaka 1000, Bangladesh.
Tel: 880-2-505769(W)
Fax: 880-2-865583(W)

Specialisation : History; International Relations.

Areas of Current Research:
Public International Law; Law of the Sea; Diplomacy.

Main Work at Hand:
War Crimes Tribunal for Former Yugoslavia: An Innovation by the UN.

Recent/Forthcoming Publications:
"Exclusive Economic Zone in the Bay of Bengal: Implications for Bangladesh & Other Littoral States", in Emajuddin Ahmed & Abdul Kalam (ed); *Bangladesh, South Asia and the World* (Academic Publishers, Dhaka); "The Bosnian Crisis: Failure of International Diplomacy", *Journal of International Relations D.U.* Vol.1, No.2, Jan-June'94.

HUQ, Saleemul (M/43)
Executive Director, Centre for Advance Studies
House 620 , Road 10A, Dhanmondi, Dhaka, Bangladesh.

Specialisation : Botany.

Areas of Current Research:
Environment; Water Resources; Nature Conversation.

Main Work at Hand:
Environment & Development in Bangladesh; Environmental Aspects of Surface Water Systems in Bangladesh.

HUSAIN, Neila (F/28)
Research Associate, BIISS, 1/46, Elephant Road
Dhaka 1000, Bangladesh.
Tel: 880-2-406234(W)
Fax: 880-2-832625(W)

Specialisation : International Relations.

Areas of Current Research:
Asia-Pacific Affairs; Bangladesh Foreign Policy; Women in Politics.

Main Work at Hand:
The Impact of Proliferation of Small Arms in South Asia: The Case of Bangladesh.

Recent/Forthcoming Publications:
"Bangladesh's Export of Manpower to Malaysia: Imperatives for Reorientation" *BIISS Journal,* Vol.17, No.2, 1996.

HUSAIN, Syed Anwar (M/48)
Dept. of History, University of Dhaka
Dhaka 1000, Bangladesh.
Fax: 880-2-817277(W)
E-mail: cdrb@ogni.com

Specialisation : British History.

Areas of Current Research:
South Asian Security; South Asian Democracy/Human Rights; South Asian Religion/Ethnicity.

Main Work at Hand:
Administration of India 1858-1924; Superpowers & Security in the Indian Ocean.

Recent/Forthcoming Publications:
Politics of Islam in Bangladesh; Bangladesh: Religion, Ethnicity & Democracy.

HUSAIN, Tabarak (M/71)
Former Foreign Secretary, No.41, Road 6A
Dhanmondi Residental Area, Dhaka, Bangladesh.

Specialisation : Economics.

Areas of Current Research:
International Affairs; Foreign Policy of Bangladesh.

Recent/Forthcoming Publications:
Articles on South Asian Affairs; SAARC; Human Rights.

HUSSAIN, Akmal (M/47)
Dept. of International Relations, University of Dhaka
Dhaka 1000, Bangladesh.
Tel: 880-2-505455(H)
Fax: 880-2-865583(W)

Specialisation : Middle-East Studies.

Areas of Current Research:
Security & Foreign Policy of the Middle East & Central Asia;
Bangladesh Foreign Policy.

Main Work at Hand:
Geopolitics of Central Asia; Security of the Persian Gulf in the Post-Cold War Era.

Recent/Forthcoming Publications:
"Emerging Pattern in the Central Asian Geopolitics", *Journal of the Asiatic Society of Bangladesh,* Vol.41, No.1, June'96.

IFTEKHARUZZAMAN, (M/45)
Executive Director
Regional Centre for Strategic Studies
4-101, BMICH, Bauddhaloka Mawatha
Colombo 7
Tel: 94-1-688601, Fax: 94-1-688602
E.mail: edrcss@sri.lanka.net

Specialisation: Economics/International Relations

Areas of current research:
Politics, Inter-State Relations and Security in South Asia; Conflict and Conflict Resolution; Development and Corporation in South Asia

Main work at hand:
South Asian Security: Multi-dimensional Concern & Approach; Bangladesh: A Weak State and Power

Recent/Forthcoming Publication:
"Institutional Sources of Conflict and Implications for Regional Cooperation in South Asia"; *Regional Economic Trends and South Asian Security* (ed); *South Asia's Security; Primacy of Internal Dimension* (ed).

ISLAM, M. Aminul (M/62)
Professor, Dept of Geography, University of Dhaka
Dhaka 1000, Bangladesh.
Tel: 880-2-505740(W) 504663(H)
Fax: 880-2-865583(W)

Specialisation : Geography.

Areas of Current Research:
Agricultural System; Resource Management; Environmental Issues.

Main Work at Hand:
Hazard Perception & Adjustment to Cyclone Hazards in the Coastal Area of Bangladesh.

Recent/Forthcoming Publications:
"Resource Management Issues in the Coastal Areas of Bangladesh", *Journal of the Asiatic Society of Bangladesh,*1992; *Environment, Land Use & Natural Hazards in Bangladesh* (University of Dhaka); Problems of Resource Management.

ISLAM, M. Nazrul (M/53)
Professor, Dept. of Political Science, University of Dhaka
Dhaka 1000, Bangladesh.
Tel: 880-2-866718(H)
Fax: 880-2-865583(W) 880-2-865583(H)
E-mail: duce@agni.com

Specialisation : Political Science.

Areas of Current Research:
Comparative Politics of South & South East Asia; Nation Building Problems of Developed & Developing Areas with Particular Emphasis on Malaysia & Pakistan.

Main Work at Hand:
Political Dimensions of Ethnic Violence in South Asia: The Case of Kashmir in India.

Recent/Forthcoming Publications:
Political & Economic Dimensions of Ethnic Pluralism in Asia and the Pacific; Pressure Groups in Bangladesh: The Case of Chamber of Commerce & Industries.

ISLAM, M. Nazrul (M/32)
Lecturer, Dept. of Geography, Jahangirnagar University
Savar, Dhaka 1342, Bangladesh.

Specialisation : Geography.

Areas of Current Research:
Fluvial Geomorphology; Hydro-Meteorology.

Main Work at Hand:
Vulnerability & Adaptive Strategies to Climate Change: A Case Study of Dhaka City.

Recent/Forthcoming Publications:
Hydrological Model Developed for South-Western Region of Bangladesh, (1995); Adjusment to Braiding Islands & Channel Width: A Study of the Jamuna River, 1996.

ISLAM, Nazrul (M/54)
Dept. of Geography, University of Dhaka
Dhaka 1000, Bangladesh.
Tel: 880-2-505740(W) 861500(H)
Fax: 880-2-865583(W)

Specialisation : Geography.

Areas of Current Research:
Urbanization & Urban Governance; Development; Environment.

Main Work at Hand:
Urban Governance; Urban Poverty.

Recent/Forthcoming Publications:
Dhaka: From City to Mega City (Dhaka University, 1996); *Urban Research in Bangladesh* (ed. C.U.S., 1994); *Urban Poor in Bangladesh* (ed. C.U.S., 1996).

ISLAM, Shamsul (M/29)
Research Associate, BIISS, 1/46 Elephant Road
Dhaka 1000, Bangladesh.

Specialisation : Mass Communication & Journalism.

Areas of Current Research:
International Communication; Role & Impact of Media in International Relations; Foreign Policy & Security Issues of South Asia, West Asia & Africa.

Main Work at Hand:
The Emergence of Taliban Factor in Afghan Politics.

Recent/Forthcoming Publications:
Recent Trends in International Communication: A Third World Perspective; Mass Media & Democratisation in Bangladesh: Experience of Nineties.

KABIR, Mohammad Humayun (M/40)
Senior Research Fellow, BIISS, 1/46, Elephant Road
Dhaka 1000, Bangladesh.
Tel: 880-2-406234(W) 880-2-819710(H)
Fax: 880-2-832625(W)
E-mail: biiss@drik.bgd.toolnet.org

Specialisation : International Law & Relations.

Areas of Current Research:
Development & Security; Problem of Governance in South Asia; Conflict Resolution in South Asia.

Main Work at Hand:
Governance, Development & National Security of Bangladesh; The Ethnic Issues in Sri Lanka: Prospects for Conflict Resolution.

Recent/Forthcoming Publications:
The India Factor in Sri Lanka's Foreign & Security Policy, 1948-1994; The Indian Ocean Rim Initiative: Bangladesh's Interest & Role; UN Peace Keeping & Bangladesh.

KARIM, Shahnaz (F/25)
Centre for Policy Dialogue, 6A, Eskaton Garden
Ramna, Dhaka 1000, Bangladesh.
Tel: 880-2-836737(W) 880-2-380427(H)
Fax: 880-2-835701(W)
E-mail: rs.cpd@drik.bgd.toolnet.org

Specialisation : International Relations.

Areas of Current Research:
The Governance of Asia; Security of Small States; Human Rights.

Recent/Forthcoming Publications:
Security Imperatives and Economic Benefits of Transit Facility: A Bangladesh Perspective.

KHAIR, Sumaiya (F/32)
Department of Law, University of Dhaka,
Dhaka 1000, Bangladesh.
Tel: 880-2-812186; 814001(H)

Specialisation : Law.

Areas of Current Research:
Gender Studies; Children's Rights; Human Rights.

Main Work at Hand:
Child Labour & Child Rights; Women's Issues.

KHAN, Abdur Rob (M/41)
BIISS, 1/46, Elephant Road
Dhaka 1000, Bangladesh.
Tel: 880-2-412693(W)
Fax: 880-2-833625(W)
E-mail: biiss@drik.bgd.toolnet.org

Specialisation : International Relations.

Areas of Current Research:
Conflict Studies; South Asian Politics; Bangladesh Domestic Politics.

Main Work at Hand:
Kashmir: Route to Interactability; Trade & Investment Links in South Asia:
Role of Japan.

Recent/Forthcoming Publications:
"Bangladesh-India Political Relations" in S.R.Chakravarty (ed.), *Bangladesh Foreign Policy*, (1994); "An Analysis of Current Hypotheses on Future Wars", *BIISS Journal,* Vol. 17, No. 3, 1996.

KHAN, Habibur Rahman (M/27)
Research Associate, BIISS, 1/46 Elephant Road
Dhaka 1000, Bangladesh.
Tel: 880-2-406234(W)
Fax: 880-2-832625(W)
E-mail: biiss@drik.bgd.toolnet.org (W)

Specialisation : Geography.

Areas of Current Research:
Environmental Studies, Natural Resource & Integrated Coastal Zone Management; Development Issues, Planning & Domestic Politics; Foreign Policy & Food Security in South & South East Asia.

Main Work at Hand:
Desertification: It's Consequences in Bangladesh; Child Labour Controversy in Garment Industry: A Move Towards Solutions.

KHAN, Maimul Ahsan (M/40)
Associate Professor, Department of Law, University of Dhaka
Dhaka 1000, Bangladesh.

Specialisation : Law.

Areas of Current Research:
Human Rights & Nationality Problems; Constitutional Law & Religious Issues; International Institutions.

Main Work at Hand:
Serb War Criminals; Human Rights Isolations; Chechen Syndrome in Russia.

Recent/Forthcoming Publications:
Environment & Human Rights: National & International Perspective; The Former Soviet Republics & Nationality Policies; Islamic Fundamentalism in the Middle Eastern Countries.

KHAN, Mizanur R. (M/41)
Senior Research Fellow, Bangladesh Institute of International & Strategic Studies, 1/46, Elephant Road, Dhaka 1000, Bangladesh.
Tel: 880-2-406234(W)
Fax: 880-2-832625(W)

Specialisation : Environmental Policy.

Areas of Current Research:
Sustainable & Participatory Development; Community/Social Forestry; Energy, Environment & Security Issues.

Main Work at Hand:
Social Dimensions of Sustainable Development: An Inquiry into the Frestry Sectors of Bangladesh & West Bengal, India.

KHAN, Mohammed Mohabbat (M/46)
Dept. of Public Administration, University of Dhaka
Dhaka 1000, Bangladesh.
Tel: 880-2-505899(W) 880-2-861411(H)
Fax: 880-2-865583(W)
E-mail: cdrb@agni.com

Specialisation : Public Administration

Areas of Current Research:
Governance; Democracy; Public Sector Reform.

Main Work at Hand:
Bureaucracy & Reform in Bangladesh; Urban Governance in Bangladesh & Pakistan.

Recent/Forthcoming Publications:
The Bangladesh Economy in Transition (New Delhi: Oxford, 1996); "Process of Democratization in Bangladesh", *Contemporary South Asia,* Vol.5, No. 2 , July 1996; "Management of an Asian Mega City: Dhaka, *The Philippine Journal of Public Administration* , Vol.39 , No.3, July 1995.

LATIF, Abdul (M/40)
Associate Professor, Dept. of Government & Politics
Jahangirnagar University, Savar, Dhaka, Bangladesh.
Tel: 416072(H)

Specialisation : International Relations.

Areas of Current Research:
Military in Politics; Security Studies; Bangladesh Politics.

Main Work at Hand:
Changing Patterns of Civil Military Relations in Bangladesh (1971-96);
Ruhingya Issue in Bangladesh: Implications for South Asia.

Recent/Forthcoming Publications:
Military Politics in Bangladesh: Ziaur Rahman Regime (1975-81); An Introduction to international Politics; Rise & Fall of a Dictator: A Study of Ershad Regime (81-90) in Bangladesh.

MAHBUB, A.Q.M. (M/43)
Dept. of Geography, University of Dhaka
Dhaka 1000, Bangladesh.
Tel: 880-2-505740 Ext.32(W)

Specialisation : Geography, Migration.

Areas of Current Research:
Population Dynamics; Urbanization; Environment.

Main Work at Hand:
Population Atlas of Bangladesh; Economic Geography of Bangladesh.

Recent/Forthcoming Publications:
Population Circulation (Migration) in Bangladesh; Slums of Dhaka City (1991); Labour Migration from Bangladesh to Asian Countries (1994).

MAHMOOD, Raisul Awal (M/43)
Chief, Population & Studies Div.,BIDS, E-17, Agargaon
Sher-e-Bangla Nagar, Dhaka 1117, Bangladesh.
Tel: 880-2-316655(W) 861501(H)
Fax: 880-2-813023(W) 862192(H)

Specialisation : Economics.

Areas of Current Research:
International Migration, Trade & Development; Population & Development Interface; Human Resource Development.

Main Work at Hand:
Dynamics of Emigration Pressure in the Context of Bangladesh; Social Clause in Trade Agreements & Implications for Trade, Human Resource Development.

Recent/Forthcoming Publications:
Analysis of Present & Future Emigration Dynamics in Bangladesh;
Globalisation with Equity: Policies for Growth in Bangladesh; Cross-
National Labour Migration & Local Level Planning.

MAHMUD, Simeen (F/45)
BIDS, E-17 Agargaon,
Sher-e-Bangla Nagar, Dhaka 1117, GPO 3854, Bangladesh.

Specialisation : Demography.

Areas of Current Research:
Women's Status & Fertility; Women's Labour Force Participation; Gender
Issues in Agriculture & Access to Resources.

Main Work at Hand:
Women's Labour Force Participation.

Recent/Forthcoming Publications:
From Women's Status to Empowerment: The Shift in the Population; Impact
of Credit Programmes on Fertility Regulating Behaviour; Women's Work in
Urban Bangladesh; Is There an Economic Rationale?

MAHTAB, Nazmunnessa (F/49)
Dept. of Public Administration, University of Dhaka
Dhaka 1000, Bangladesh.
Tel: 880-2-841445(H)
Fax: 880-2-832749(H)

Specialisation : Rural Development.

Areas of Current Research:
Women in Development; Development Administration; Rural Development
& Local Government.

Main Work at Hand:
Participation: of Women at a Local Level; Impact/Evaluation Study of
NGO's on Women's Development.

Recent/Forthcoming Publications:
Participation of Women at Local Level; NGO's & Women's Development;
Girl Child: Victims of Poverty & Violence.

MANIRUZZAMAN, Talukder (M/56)
Professor, Dept. of Political Science, Dhaka University
Dhaka 1000, Bangladesh.
Tel: 880-2-503083(W) 506311(H)
Fax: 880-2-865583(W)

Specialisation : Political Science.

Areas of Current Research:
Politics of Developing Areas; Defence & Military; Security of Small States.

Main Work at Hand:
Effects of Arms Transfer in Developing Countries; Defence Policy of Japan After 2nd World War.

Recent/Forthcoming Publications:
Military Withdrawal from Politics (Bellinger, USA); *Politics & Security of Bangladesh* (UPL,1995); Defence Policy of Japan.

MIRZA, M. Hussan (M/33)
Centre for Policy Dialogue, 6A Eskaton Garden
Dhaka 1000, Bangladesh.
Tel: 880-2-837055(W) 839880(H)

Specialisation : Public Administration; Urban & Development Studies.

Areas of Current Research:
Governance & Development; Human Rights; Gender & Population Studies.

MONSOOR, Taslima (F/36)
Assistant Professor, Department of Law, University of Dhaka
Dhaka 1000, Bangladesh.
Tel: 880-2-314459(H)

Specialisation : Family Law; Women.

Areas of Current Research:
Women; Family Law.

Main Work at Hand:
Family Law & it's Impact on Women in Bangladesh.

Recent/Forthcoming Publications:
Impact of Family Law on Women in Bangladesh; Family Courts & Women: A Step Towards Right Direction?

MORSHED, A.K.H. (M/62)
125, Gulshan Avenue
Gulshan Model Town
Dhaka, Bangladesh.

Specialisation : International Law.

Areas of Current Research:
Public International Law; Law of Non-Navigational Uses of International Water; Law of the Sea.

MOSHAREF, Mir Ali (M/28)
Lecturer, Dept. of Geography, University of Dhaka,
Dhaka 1000, Bangladesh.
Tel: 880-2-505740(W)
Fax: 880-2-865583(W)

Specialisation : Geography.

Areas of Current Research:
Fluvial Geomorphology; Environmental Studies & Natural Hazard.

Main Work at Hand:
Sedimentalogical Study of Ganges Delta in Bangladesh; Flood of Bangladesh.

Recent/Forthcoming Publications:
"Alluvial Characteristics of the Ganges Floodplain", *The Journal of BNGA,*
Vol-18, 1993 (Bangladesh); Morphological Characteristics of Ganges
Floodplain: A Study of Moribund Delta, *Bhugol Patrika,* No.12/1993.

MURSHED, Yasmeen (F/50)
Centre for Analysis & Choice, House 65 , Road 6/A (new)
Dhanmondi, Dhaka 1209, Bangladesh.
Tel: 880-2-886858(W) 880-2-883554(H)
E-mail: murshed.@driktap.tool.nl

Specialisation : Economics.

Areas of Current Research:
Democratic Development; Women in Politics.

Main Work at Hand:
Skills Development, Training for Women Political Candidates.

NAZEM, Nurul Islam (M/39)
Associate Professor, Dept. of Geography, Jahangirnagar University
Savar, Dhaka 1342, Bangladesh.

Specialisation : Rural-Urban Relations.

Areas of Current Research:
Urban & Regional Development; Resources & Environmental Management;
Geopolitics in South Asia.

Main Work at Hand:
Urban Poverty in Bangladesh; Strategic Role of Small Towns in Rural
Development of Bangladesh.

Recent/Forthcoming Publications:
"Urbanization & Rural Development in Bangladesh", *Geographical Review*
(Oxford) 8(1) 1994; "The Impact of River Control on an International

Boundary", *World Boundaries*, Vol.3, (Eurasia); Urban Poor in Bangladesh.

NURUZZAMAN, Md. (M/29)
Lecturer, Dept. of International Relations
University of Dhaka, Dhaka 1000, Bangladesh.
Tel: 880-2-505769(W)
Fax: 880-2-865583(W)

Specialisation : International Relations.

Areas of Current Research:
Conflict Analysis & Peace Studies, South Asian Strategic Studies.

Main Work at Hand:
Nuclear Proliferation in South Asia: The Peace Options of the Non-Nuclear Smaller South Asian States

Recent/Forthcoming Publications:
"UN Intervention on Humanitarian Grounds: New Humanism or a Chaotic Political Doctrine?", *Journal of International Relations,* D.U.

RAHMAN, A. Aminur (M/45)
Professor, Dept. of Political Science, University of Dhaka
Dhaka 1000, Bangladesh.
Tel: 880-2-505809(W) 507338(H)
Fax: 880-2-865583(H)

Specialisation : Local Government.

Areas of Current Research:
Social Change; Comparative Politics.

Main Work at Hand:
Socio-Cultural & Economic Aspect of Tribal People in the Chittagong Hill Tracts (C.H.T.).

RAHMAN, Ataur (M/47)
Professor, Dept. of Political Science, University of Dhaka,
Dhaka 1000, Bangladesh.

Specialisation : Political Science.

Areas of Current Research:
East Asian Strategic Framework; South Asian Politics; Governance.

Main Work at Hand:
Governance in South Asia.

Recent/Forthcoming Publications:
"Towards a New Strategic Framework in East Asia", *BIISS Journal*, 1996.

RAHMAN, Mohammad Anisur (M/26)
Centre for Policy Dialogue, 6A Eskaton Garden
Ramna, Dhaka 1000, Bangladesh.
Tel: 880-2-837055(W) 381113(H)
Fax: 880-2-835701(W)
E-mail: rs.cpd@drik.bgd.toolnet.org

Specialisation : Economics

Areas of Current Research:
International Trade; Impact of Regional Co-Operation & Integration; Development Economics.

Main Work at Hand:
US-Japan Relations: A Strategic Trade Policy Analysis.

Recent/Forthcoming Publications:
"Harkin's Bill & the Issue of Child Labour; US-Japan Relations", *Economic Observer*, May 1996.

RAHMAN, Mustafizur (M/40)
Co-ordinator, Independent Review Project, Centre for Policy Dialogue
6A Eskaton Garden, Ramna, Dhaka 1000, Bangladesh.
Tel: 880-2-837055(W)
Fax: 880-2-835701(W)
E-mail: rs.cpd@drik.bgd.toolnet.org

Specialisation : Development Economics.

Areas of Current Research:
Trade Economics; Development Economics; Credit Policy.

Main Work at Hand:
Regional Co-operation in Trade Among South Asian Countries; Export Incentives in Bangladesh.

Recent/Forthcoming Publications:
"G.S.P. & GATT", *Journal of Buisness Studies*, D.U. July, 1995; "Export Incentives in a Comparative Environment", *Journal of Buisness Administration*, D.U. 1995; *Regional Co-operation Among South Asian Countries* (to be Published by Tata McGraw Hill, India).

RAHMAN, Shamsur (M/31)
BIISS, 1/46, Elephant Road
Dhaka 1000, Bangladesh.
Tel: 880-2-406234(W) 404031(H)
Fax: 880-2-832625(W)

Specialisation : Economics.

Areas of Current Research:
Macro-Economic Policy Issues; Environment, Gender & Poverty Issues;
Trade & Agricultural Issues.

Main Work at Hand:
Co-operation Schemes Between SAARC & Japan; Indo-Bangladesh Trade
Issues.

Recent/Forthcoming Publications:
Structural Adjustment Programme in South-Asian Countries: Performance of
Macro-Economic Indicators; Towards an Integrated Environmental Policy for
South Asia.

RAHMAN, Tareque Shamsur (M/40)
Dept. of Govt. & Politics, Jahangirnagar University,
Savar, Dhaka 1000, Bangladesh.

Specialisation : International Relations.

Areas of Current Research:
International Politics; Soviet/Russian Foreign Policy.

Main Work at Hand:
Bangladesh: Politics of Last 25 Years; International Law & Farakka Issue.

Recent/Forthcoming Publications:
Bangladesh: Society & Politics (Dhaka, 1992).

RASHEED, K.B. Sajjadur (M/55)
Department of Geography, University of Dhaka
Dhaka 1000, Bangladesh.
Tel: 880-2-862413(H)
Fax: 880-2-865583(W)

Specialisation : Geography.

Areas of Current Research:
Environment; Water Resources; Rural Development.

Main Work at Hand:
Vulnerability & Adaptation Strategies in Riverine Flooding.

Recent/Forthcoming Publications:
Nepal's Water Resources, in G.D. Chapman & M. Thompson (eds); *Water &
the Quest for Sustainable Development in the Ganges Valley* (1995, London).

RAZZAQUE, Mohammad Abdur (M/24)
Research Associate, Centre for Policy Dialogue
6A, Eskaton Garden, Ramna, Dhaka 1000, Bangladesh.
Tel: 880-2-837055(W)
Fax: 880-2-835701(W)
E-mail: rs.cpd@drik.6gd.toolnet.org

Specialisation : Economics.

Areas of Current Research:
International Economics; Development Economics; Economic Co-operation
& Regional Integration.

Main Work at Hand:
US-Japan Relations: A Strategic Trade Policy Analysis; Quality of Public
Expenditure in Bangladesh.

Recent/Forthcoming Publications:
"Harkin's Bill & the Issue of Child Labour; US-Japan Relation", *Economic
Observer*, May 1996.

RIZVI, Gowher (M/47)
Representative, 55, Lodi Estate
New Delhi 110003, India.
Tel: 91-11-4619441(W)
Fax: 91-11-4627147(W)

Specialisation : History.

Areas of Current Research:
International Relations & Security of South & South West Asia; Politics of
Developing Society; Political Islam/Impact on International Society.

Main Work at Hand:
Governance, Democracy & Civil Society in South Asia; Self Induced De-
pendence: A Study of the Post Colonial States.

Recent/Forthcoming Publications:
South Asia in a Changing World Order, 1994; *South Asian Insecurity &
Great Powers*, 1988; *The Struggle for the Restoration of Democracy in
Bangladesh*, 1988.

SABUR, A.K.M. Abdul (M/39)
Senior Research Fellow, BIISS, 1/46, Elephant Road
Dhaka 1000, Bangladesh.
Tel: 880-2-406234(W)
Fax: 880-2-832625(W)
E-mail: biiss@drik.bgd.toolnet.org (W)

Specialisation : International Relations.

Areas of Current Research :
Domestic Policies & Inter State Relations in South Asia; Problems of Third World Security with Particular Reference to South Asia.

Main Work at Hand:
Indo-US Relations in the Post-Cold War Era.

Recent/Forthcoming Publications:
"Challenges of Governance in India: Fundamentals Under Threats", *BIISS Papers*, No.15, July 1995; "South Asian Security in the Post-Cold War Era: Issues & Outlook", *Contemporary South Asia*, Vol. 3, No. 2, 1994.

SEN, Binayak (M/37)
BIDS, E-17 Agargaon, Sher-e-Bangla Nagar
Dhaka 1207, Bangladesh.
Tel: 880-2-813623(W)

Specialisation : Economics.

Areas of Current Research:
Poverty & Income Distribution; Rural Economics; Economic & Social History.

Recent/Forthcoming Publications:
"Changes in Poverty in a Cross-Section Countries", In A.A. Abdullah & A.R, Khan (eds). *State, Market & Development Essays in Honour of Rehman Sobhan*, (University Press Ltd. Dhaka, 1996); "Monitoring Progress in Poverty in Bangladesh", in *Economic Development & Cultural Change*; *Social Dimension of Adjustment, World Bank Experience, 1980-1993* (Operation Evaluation Department, World Bank, 1996); "Rural Poverty in Bangladesh Trends & Determinanats", *Asian Development Review*, Vol.10, No.1.

SHAH, Muhammad Azhar Zafar (M/39)
Dept. of International Relations, University of Dhaka
Dhaka 1000, Bangladesh.
Tel: 880-2-9558368(W)

Specialisation : International Relations.

Areas of Current Research:
South Asia; Middle East; East Europe.

Recent/Forthcoming Publications:
India and the Super-Powers (University Press Ltd., Dhaka,1983).

SHAHIDUZZAMAN, Muhammad (M/42)
Professor, Dept. of International Relations
University of Dhaka, Dhaka 1000, Bangladesh.
Tel: 880-2-316644(H)

Specialisation : International Relations.

Areas of Current Research:
International Security: Foreign Policy Analysis; Insurgency & Ethnic Crisis: American Foreign Policy.

Main Work at Hand:
Nuclear Proliferation & South Asian Security; Military Strategy of Bangladesh.

Recent/Forthcoming Publications:
"Alliance Reliability in the Post-Cold War Context & Bangladesh Military Strategy", *Journal of the Asiatic Society of Bangladesh*, Dec 1995; "The State and Nation Approach in Resolving Problems of Ethnicity", *BIISS Journal,* July 1995; Indian Foreign Policy Transformation in the Nineties.

SHAMIM, Ishrat (F/48)
Dept. of Sociology, University of Dhaka,
Dhaka 1000, Bangladesh.

Specialisation : Applied Sociology; Anthropology.

Areas of Current Research:
Violence Against Women; Women & Environment; Women in Poverty Situation.

Main Work at Hand:
Trafficking Women & Children; Domestic Violence and Women.

Recent/Forthcoming Publications:
Energy & Water Crisis in Rural Households: Linkages with Women's Work & Time; Rural Women in Poverty: NGO Interventions; Status of Widows in Bangladesh: Issues and Consensus.

SHEIKH, Yunus Ali (M/30)
Research Associate, BIISS, 1/46 Elephant Road
Dhaka 1000, Bangladesh.
Tel: 880-2-418262(H)

Specialisation : Mass Communication & Journalism.

Areas of Current Research:
Role & Impact of Media in International Relations; International Communication; Foreign Policy & Security Issues of South Asia & West Asia.

Recent/Forthcoming Publications:
Bangladesh Foreign Policy Dilemma as Reflected in the Media During Gulf War 1990-91.

SHELLEY, Mizanur Rahman (M/52)
Chairman, Centre for Development Research
55 , Dhanmondi R/A,Road 8-A
G.P.O. Box 4070, Dhaka 1209, Bangladesh.
Tel: 880-2-811877(W)
Fax: 880-2-817277(W)
E-mail: cdrb@agni.com

Specialisation : International Politics.

Areas of Current Research:
Socio-Economic & Rural Development; Resource Management, Community Foresty; Regional & International Co-operation and Development.

Main Work at Hand:
Development Institutions; Education, including Non-Formal Education; Socio-Economic Development including Women, Youth & Disadvantaged Groups.

Recent/Forthcoming Publications:
Privatising Industrial Regulatory Functions in Bangladesh; *The Chittagong Hill Tracts of Bangladesh: The Untold Story; Foreign Policy in a Changing World; Bangladesh: Continuity & Change.*

ULLAH, A.K.M. Shahid (M/55)
Professor, Department of Political Science, University of Dhaka
Dhaka 1000, Bangladesh.

Specialisation : Political Science.

Areas of Current Research:
Electoral Politics.

Main Work at Hand:
Parliamentry Election of Bangladesh-1996.

ZAHID, Anowar (M/29)
Assistant Lecturer, Department of Law
University of Dhaka, Dhaka 1000
Bangladesh.
Tel: 880-2-815954(H)
Fax: 880-2-815954(W)

Specialisation : Law.

Areas of Current Research:
Refugees in Asia particularly Rohingya Refugees; Law & Policy in South Asia; Islam in South Asia.

Main Work at Hand:
Refugees, Law & State: A Case Study of Rohingya Refugees in Bangladesh.

Recent/Forthcoming Publications:
Fundamental Rights & their Enforcement in Bangladesh; Preventive Detention & Liberty: Bangladesh as Case Study.

■ Rajshahi

ALAM, Syed Rafiqul (M/45)
Professor, Dept. of Geography, Rajshahi University
Rajshahi 6205, Bangladesh.

Specialisation : Geography.

Areas of Current Research:
Environment Studies; Landuse; Transportation.

Main Work at Hand:
"Land/Water Interface in Flood Plains of Bangladesh", A Project of ODA.

Recent/Forthcoming Publications:
"Landuse Characteristics of Khulna City & it's Impact on Environment", *The Journal of the Bangladesh National Geographical Association*, Nov. 1995, Vol. 19.

CHOWDURY, Farah Deeba (F/29)
Dept. of Political Science, University of Rajshahi
Rajshahi 6205, Bangladesh.

Specialisation : Government & Politics.

Areas of Current Research:
Women's Studies.

Main Work at Hand:
Socio-Economic Context of Child Marriage.

Recent/Forthcoming Publications:
Women in University Administration: A Case Study of Rajshahi University.

HABIB, Shah Ehsan (M/28)
Lecturer in Sociology, Department of Sociology
University of Rajshahi, Rajshahi 6205, Bangladesh.
Tel: 880-721-3041-9(W)
Fax: 880-721-2064(W)
E-mail: ru.phy@drik.bgd.toolnet.org

Specialisation : Sociology

Areas of Current Research:
Problems of Cultural Conflict in South Asia; Community Development;
Political Development.

Main Work at Hand:
Problems of Cultural Assimilation in Australia.

Recent/Forthcoming Publications:
Rapid Assessment Study: Bangladesh Drug Situation; Impact Assessment of
Proshika.

HAIDER, Zaglul (M/33)
Dept. of Political Science, University of Rajshahi
Rajshahi 6205, Bangladesh.

Areas of Current Research:
Foreign Policy; International Politics; Comparative Politics.

Main Work at Hand:
The Controversial Sixth General Election in Bangladesh & it's Aftermath.

Recent/Forthcoming Publications:
"Bangladesh-China Relations: A Review".

HASSAN, Zahidul (M/40)
Associate Professor, Dept. of Geography, University of Rajshahi
Rajshahi 6205, Bangladesh.

Specialisation : Geography.

Areas of Current Research :
Agricultural Geography; Resource Management; Environmental Studies.

Main Work at Hand:
Urban Landuse Pattern & the Environmental Characteristics of Bogra &
Khulna-A Study on Some Aspects of Urban Ecology; Agricultural Systems in
Bangladesh-A Case of GK Project Area.

Recent/Forthcoming Publications:
"Farmer Responses to Drought in Bangladesh: A Case Sudy of Villages",
JIBS, Vol. 17, 1994; "Crop-Diversity & Crop-Diversification Programme in
Bangladesh", *Bhugol Patrica*, Vol. 14, 1995; "Spatial Pattern of Agricultural
Development in Bangladesh", *JIBS*, Vol.19, 1996.

HOQUE, A. N. Shamsul (M/63)
Dept. of Political Science, Rajshahi University
Rajshahi 6205, Bangladesh.

Tel: 880-721-2355(W)
Fax: 880-721-2064(W)
E-mail: ru.phy@drik.bgd.toolnet.org

Specialisation : Public Administration.

Areas of Current Research:
Political Systems Including Bangladesh; International Politics;
Administration & Local Government.

Main Work at Hand:
Politics, Administration & Development in Bangladesh; Reorganization of
the UN.

Recent/Forthcoming Publications:
"Politics & *Bureaucracy in Bangladesh"*, Lok Raj Baral: (ed); *South Asia:
Democracy and the Road Ahead;* "Priority of American Foreign Policy
During the Cold War & After", American Studies in the Context of Democ-
racy, Economy & Environment, "Environment & Health", *Journal of the IBS*,
Vol 17, 1994.

HUQ, M. Abdul Fazal (M/61)
Professor of Political Science, Rajshahi University, Rajshahi, Bangladesh.

Specialisation : Political Science.

Areas of Current Research:
Government and Politics in Bangladesh; International Law and Organization.

Recent/Forthcoming Publications:
"The Parliament and Politics of Bangladesh", *South Asian Studies*, Jaipur
India, Vol.22, No.2, 1987; "The Problem of National Identity in Bangladesh",
The Journal of Social Studies, Dhaka, No.24, April,1984.

IMAM, Muhammad Hasan (M/39)
Associate Professor, Department of Sociology
University of Rajshahi, Rajshahi 6205, Bangladesh.
Fax: 880-721-2064(W)

Specialisation : Sociology; Rural Sociology.

Areas of Current Research:
Agriculure; Environment; Resource Management.

Main Work at Hand:
Humanist-Environmentalist Requirements for Sustainable Bangladesh;
Putting Poor People First.

Recent/Forthcoming Publications:
Political Ecology of Environmental Degradation in Bangladesh; Socio-

Ecologically Perceived Issues & the Need for Participatory Institution
Building.

ISLAM, Saiful (M/42)
Department of Economics, University of Rajshahi,
Rajshahi 6205, Bangladesh.
Tel: 880-721-3041-49(W)
Fax: 880-721-2064(W)
E-mail: ru.phy@drik.bgd.toolnet.,org

Specialisation : Macro & International Economies.

Areas of Current Research:
Foreign Direct Investment; Balance of Payments.

Main Work at Hand:
Promoting Regional Economic Coorperation in Foreign Direct Investment.

Recent/Forthcoming Publications:
"Industrial Development & the Guidance Policy Finance: The Case of the
Japanese Automobile Industry", *Asian Economic Journal*, Vol.8, No.3,
November 1994.

KHAN, Jafar Reza (M/57)
Professor, Dept of Geography, University of Rajshahi
Rajshahi 6205, Bangladesh.
Tel: 880-721-3041

Specialisation : Geography; Agricultural Land Use.

Areas of Current Research:
Environmental Geography; Land Use Planning.

Main Work at Hand:
Equity & Development, A Socio-Economic Study of Three Villages; Land
Use & Women Labour Force in Bangladesh.

Recent/Forthcoming Publications:
Environment & the Changing Pattern of Land Use: A Study in Utilization of
Agricultural Resources in Bangladesh; Environment, Development &
Landuse in the Hill Region of Bangladesh; Social Environment & Participa-
tion: of Women in Informal Economic Activities.

KHAN, Moazzem Hossain (M/40)
Associate Professor, Department of Economics
University of Rajshahi, Rajshahi 6205, Bangladesh.

Specialisation : Economics.

Areas of Current Research:
Public Sector Economics; Environmental Economics; Development Economics.

Main Work at Hand:
Economic Role of the Government; Environmental Planning.

Recent/Forthcoming Publications:
"De-Emphasizing of the Role of Government in the Economic Development of Bangladesh", *Bangladesh Journal of Political Economy*, 1995; *Environment & Afforestation: An Alternative Model for Bangladesh SSJ*, (Rajshahi University,1994);

KHATUN, Ferdousi (F/25)
M. Phil. Fellow, Dept. of History, University of Rajshahi
Rajshahi 6205, Bangladesh.

Specialisation : History.

Areas of Current Research:
Sino-Japanese Relations.

Main Work at Hand:
US Reactions to the Sino-Japanese War, 1937-1945.

MITRA, Priti Kumar (M/50)
Institute of Bangladesh Studies, Rajshahi University,
Rajshahi 6205, Bangladesh.

Specialisation : History

Areas of Current Research:
Intellectual History of South Asia; Ancient South Asia & Buddhism.

Main Work at Hand:
"Dissent of Nazrul Islam: It's History & Poetry"; A translation into Bengali of *A History of Freedom of Thought* by J.B. Bury.

Recent/Forthcoming Publications:
"Dissent in Ancient India", In Jagdish Sharma (ed.) *Ideas & Individuals in Pre-Modern India* (Kusumanjali Prakashan, Jodhpur, India); "History of the Liberation Struggle of the Bengali People", in A.F.S. Ahmed, *Bangali Muktisangramen Itihash* (in Bengali) (Bangladesh Muktijoddha Gabeshana Kendra, Dhaka 1996).

MORSHED, Golam (M/58)
Professor, Department of Political Science, Rajshahi University
Rajshahi 6205, Bangladesh.
Tel: 880-721-3041-49(W)

Specialisation : Political Science.

Areas of Current Research:
Comparative Politics; Political Theory; Political Sociology.

Main Work at Hand:
Parliamentary Democracy in South Asia.

Recent/Forthcoming Publications:
East Bengal Provincial Elections of 1954 & Germination of Separatist Movement in East Pakistan; Pakistan's 1970 Elections & the Liberation of Bangladesh: A Political Analysis; Ayub Khan's Autocratic Regime & East Pakistani Regionalism.

NOMAN, A.N.K. (M/26)
Lecturer, Dept of Economics, University of Rajshahi
Rajshahi 6205, Bangladesh.
Tel: 880-721-3041-9(W)
Fax: 880-721-2064(W)
E-mail: pu.phy@drik.bgd.toolnet.org

Specialisation : Economics.

Areas of Current Research:
Economic Integration in South & South-East Asia; Effect of Open Market Economic Policy on Poverty Alleviation & Employment Generation in South Asian Countries; Role of Government in Economic Development of Third World Countries.

QUASEM, Abdul (M/45)
Chairman, Dept of Political Science, University of Rajshahi
Rajshahi 6205, Bangladesh.
Tel: 880-3041-9 Ext.-391(W) .

Areas of Current Research:
Politics of Development.

Main Work at Hand:
Comparative Politics; Modern Political Economy.

RAHMAN, M. Habibur (M/49)
Professor of Law, Rajshahi University
Rajshahi 6205, Bangladesh.
Fax: 880-721-2064(W)
E-mail: ru.phy@drik.bgd.toolnet.org

Specialisation : International Law.

Areas of Current Research:
Law of the Sea; Environment; Foreign Policy of Bangladesh.

Main Work at Hand:
Human Rights Under Democracy: Vulnerability Evaluation; Foreign Policy of Bangladesh.

Recent/Forthcoming Publications:
"Human Rights Under Bangladesh Constitutions: Pro and Anti National and International Perceptions", *South Asian Studies*; The Role of Law & Technology for Maintaining Environment.

RAHMAN, Muhammad Taufiqur (M/23)
Dept. of Political Science, University of Rajshahi,
Rajshahi 6205, Bangladesh.
Tel: 880-721-3041/453(W)
Fax: 880-721-2064(W)

Specialisation : Government & Politics.

Areas of Current Research:
Foreign Policy of Bangladesh; South Asian Politics; Bangladesh & the Muslim World.

Main Work at Hand:
Governmental System in Bangladesh & Sri Lanka After 1971; Bangladesh & South Asian Cooperation.

REZA, Salim (M/27)
Lecturer, Dept. of Mass Communication, Rajshahi University
Rajshahi 6205, Bangladesh.

Specialisation : Mass Communication and Journalism.

Areas of Current Research:
Communication & Religion in South Asia; Mass Media & Image of the Countries in South Asia; Communication & Politics in South Asia.

Main Work at Hand:
A Study of the Jumma Speech: Communication from the Mosques; Indo-Bangladesh Relations.

ZAMAN, Faruk-uz (M/40)
Associate Professor of History,
University of Rajshahi,
Rajshahi 6205, Bangladesh.
Fax: 880-721-2064(W)

Specialisation : History; US History.

Areas of Current Research:
International Relations; US-South Asia Relations; History of the Freedom Movement of Bangladesh.

Main Work at Hand:
US & the Independence Movement in Bangladesh: A Case Study of Super Power Intervention in a Regional Conflict; Genesis of the Historic Six-Point Movement in Bangladesh.

Recent/Forthcoming Publications:
"The Provincial Elections of 1937: A Turning Point in Indian History", *Rajshahi University Studies*, Part A, Vol. XVII, 1989; "Yet Another Way to Know Jefferson", *Rajshahi University Studies*, Part A, Vol. 21-22, 1995; Post Cold War US Policy Towards South Asia.

ZAMAN, Nasima (F/41)
Dept. of Political Science, University of Rajshahi
Rajshahi 6205, Bangladesh.
Tel: 880-721-3041-9 Ext.391(W) 325367(H)

Specialisation : Political Science.

Areas of Current Research:
Political Development in South Asia; Regional Arrangements.

Main Work at Hand:
France-Bangladesh Relations: The Changing Image; Kashmir - the Destabilizing Factor in South Asia.

TSERING, Dechen (F)
Programme Officer, Planning and Policy Division
National Environment Commission, Royal Government of Bhutan
Thimphu, Bhutan.
Tel: 975-2-23384, 24323
Fax: 975-2-23385

Specialisation: Environment & Development

Areas of Current Research:
Environment and National Resource Management of Bhutan.

Recent/Forthcoming Publictions:
Bhutanese Experience of Sharing and Management of National Resources.

TSHONG, Gonam (M/39)
Bhutan Broadcasting Service, P.O. Box 101
Thimphu, Bhutan.
Tel: 975-2-23580(W)
Fax: 975-2-23073(W)

Specialisation : Development Studies.

Areas of Current Research:
Telecommunication & Mass Media; Political Economy; Regional Alliances & Treaties of Cooperation, Security, Trade etc.

CANADA

DELVOIE, Louis A. (M/56)
Centre for International Relations, Queen's University
Kingston, Ontario, Canada K7L 3N6.
Tel: 1-613-545-2383(W)
Fax: 1-613-545-6885(W)

Specialisation : History.

Areas of Current Research:
Canadian Foreign / Defence Policy; Conflict & Security in South Asia; International Relations with the Maghreb.

Main Work at Hand:
Afghanistan: A Failed State of the 1990s; Canada's Relations with the Maghreb.

Recent/Forthcoming Publications:
Hesitant Engagement: Canada & South Asian Security, (Kingston: QCIR, 1995); "The Islamization of Pakistan's Foreign Policy", in *International Journal*, Toronto, 1995-96.

LATHAN, Andrew (M/32)
York Centre for International & Security Studies, York Lanes
York University, 4700 Keele Street, North York, Ontario M3J 1P3,Canada.

Specialisation : International Relations.

Areas of Current Research:
Light Weapon Proliferation; South Asian Security; Political Economy of Arms Production & Transfers.

Main Work at Hand:
Cooperative Security in South Asia, a Concept Paper.

CHINA

SHEN, Dingli (M/35)
Centre for American Studies, Fudan University
220, Handan Road, Shanghai 200433, China.
Tel: 86-21-6549-2222(W)
Fax: 86-21-6548-8949(W)
E-mail: dlshen@fudan.ac.cn

Specialisation : Physics.

Areas of Current Research:
Arms Control & Regional Security; Nuclear Arms Control & Nonproliferation; China-US Relations.

Main Work at Hand:
Shanghai Initiative; South Asian Summer Workshops on Security, Technology & Arms Control.

Recent/Forthcoming Publications:
China & South Asian Relations in the 1990's, Co-ed, (Sichuan People's Press, Chengdu, 1995); Remarks on Issues of Indo-Pakistani & Sino-Indian Confidence Building (in Chinese & English), *Shanghai Initiative*, published by UN.

YAN, Wenhua (F/26)
Institute for International Studies, East China Normal University
200062 Shanghai, China.
Tel: 86-21-6254-9721(W)
Fax: 86-21-6245-1966(W)

Specialisation : International Relations.

Areas of Current Research:
International Relations in Southeast Asia; The Security of South Asia.

Main Work at Hand:
Economy, Politics & International Relations of the Current World; China & it's Southern Neighbours.

Recent/Forthcoming Publications:
"The NATO Eastern Expansion & Central Europe", *Eastern Europe & Central Asia Today*, Jan 1996; "Enemies are Bound to Meet: Vietnam & the USA.", *International Survey,* Jan 1995.

DENMARK

MADSEN, Stig Toft (M/46)
Nordic Institute of Asian Studies, Leifsgade 33
DK-2300 Copenhagen S., Denmark.
Tel: 45-3-1541524(W)
E-mail: stigonias.ku.dk

Specialisation : Anthropology; Sociology.

Main Work at Hand:
Human Rights; The Environment; South Asian History.

Recent/Forthcoming Publications:
State, Society & Human Rights in South Asia, (Manohar, New Delhi, 1996).

JAFFRLOT, Christophe (M/31)
Ceri - MSH, 54 Bd Raspail
75007 Paris, France.

Specialisation : Political Science.

Areas of Current Research:
South Asia.

Main Work at Hand:
The Hindu Nationalist Movement & Indian Politics.

Recent/Forthcoming Publications:
L'Inde Contempozaive, (Paris, Fayard, 1996.)

MEYER, Eric (M/53)
INALCO - Dauphine, 2 Place De Lattre
75016 Paris, France.
Fax: 33-433-9489(W)
E-mail: emyer@inalco.fr

Specialisation : History.

Areas of Current Research:
Modern History of South Asia; Sri Lanka.

Recent/Forthcoming Publications:
"Enclave Plantations Hemmed in Villages & Dualistic Representations in Colonial Ceylon", *Journal of Peasant Studies 19,* 1996; "On the Concepts of Violence & Nonviolence in Hinduism & Indian Society", *South Asia Research 14,* 1994; "The Specificity of Sri Lanka, Towards a Comparative Modern History of Sri Lanka & India", *Economic & Political Weekly,* Mumbai, 17.02.1996, *& Lanka Guardian,* 15.03.1996.

KLOTZ, Sabine (F/34)
South Asia Institute, University of Heidelberg, Germany
IM Nevenheimer Feld 330, D-69120 Heidelberg, Germany.
Tel: 6221-546189(W) 6221-28742(H)
Fax: 6221-544591(H)

Specialisation : Political Science.

Areas of Current Research:
International Relations; Conflict Research; South Asian Studies.

Main Work at Hand:
Peaceful Settlement of Conflicts within Societies.

Recent/Forthcoming Publications:
"Regional Conflicts in South & Southeast Asia", in: Pfetschi (ed.), *Konflikte Seit* 1945, 1991-1996; "Peaceful Conflict Settlement", in, Birckenbach / Wellmanni / Jaeseried (ed.), Jahrbuch Frieden 1996 & 1997, 1995-1996.

MAASS, Citha D. (F/49)
Stiftung Wissenschaft Und Politik (SWP)
Research Institute for International Affairs,
Zeller Weg 27, D-82067 Ebenhausen, Germany.
Tel: 49-8161-63425(H)

Specialisation : Political Science; Economics; European History

Main Work at Hand:
India, Nepal, Sri Lanka; South-South Relations.

Recent/Forthcoming Publications:
"Reorientation of Indian Foreign Policy After the Cold War", *Aussenpolitik,* Vol.44, No.1, 1993; "Kashmir Conflict: Confidence Building Measures & Finding Solutions, Comments from a German Point of View", K.F. Yusuf (ed.) Perspectives on Kashmir, *Pakistan Forum,* Islamabad 1994; An Analysis of India's & Pakistan's Rejection of the Nuclear Nonproliferation Treaty (NPT).

WAGNER, Christian E. (M/37)
University of Mainz, Dept. of Political Sciences, Abt. Mols,
Colonel-Kleinman-Weg 2, 55099 Mainz, Germany.
Tel: 49-61-31392728(W)
Fax: 49-61-31393328(W)
E-mail: wagnerc@verwaltung.unimainz.de

Specialisation : Political Science.

Areas of Current Research:
Regional Cooperation in Comparative Perspective; Democratization; Nationalism & Ethnic Conflict.

Main Work at Hand:
Regional Cooperation in Europe & South Asia, Regional Conflicts, Arms Race & Nuclear Proliferation in South Asia.

Recent/Forthcoming Publications:
Regional Economic Trends & South Asian Security - A European Perspective; Democratization in Nepal; ASEAN & SAARC in Comparative Perspective, Mainz 1994.

ZINGEL, Wolfgang Peter (M)
South Asia Institute, 1M Neuenheimer Feld 330,
D-69120 Heidelberg, Germany.

Specialisation : Economics.

Areas of Current Research:
Economics, South Asia: Agriculture, Development, International, Regional.

Recent/Forthcoming Publications:
Alleviating Urban Poverty - The Palestini Way; Regional Sustainable Development? South Asia Joining Hands.

INDIA

■ *Aligarh (UP)*

NIZAMI, Taufiq Ahmad (M/50)
Dept. of Political Science, Aligarh Muslim University
Aligarh 202002, India.
Tel: 401720(W) 403183(H)
Fax: 401617(W)

Specialisation : International Relations.

Areas of Current Research:
International Relations including theoretical aspects; South Asia and Superpowers; Indian Foreign Policy.

Main Work at Hand:
US Perceptions of South Asia; Jawarharlal Nehru's Contribution to International Politics.

Recent/Forthcoming Publications:
The Communist Party & India's Foreign Policy, Barnes & Noble USA; The
Divided Left; Tanzania & the World.

■ Allahabad (UP)

KHANNA, D.D. (M/72)
58, Balrampur house, Allahabad 211002
India.
Tel: 642558(H)
Fax: 640211(H)

Specialisation : Defence & Strategic Studies.

Areas of Current Research:
South Asian Security Problems; Global Disarmament & Arms Control;
India-Pakistan.

Recent/Forthcoming Publications:
Dialogue of the Deaf - India, Pakistan Divide 1990; *Defence Vs Development*, (Konark Publications, 1993). *Sustainable Development, Environmental Security & Disarmament Interface in South Asia*, (Macmillans,1997).

■ Almora (UP)

FAROOQUEE, Nehal Ahmed (M/32)
G.B. Pant Institute of Himalayan Environment & Development Studies
Kosi, Almora 263643, India.
Tel: 5962-81144(W)

Specialisation : Political Science.

Areas of Current Research:
International Relations; Environment & Development; Cultural Studies.

Main Work at Hand:
Resource Use Pattern of High Attitude Himalayan Society; Tribal Perception
of Development.

Recent/Forthcoming Publications:
"Conservation & Utilization of Medicinal Plants in High Hills of Central
Himalaya", *Environmental Conservation* 1996; "Diversity, an Important
Componant of Himalayan Sustainability", *Man in India* 1996.

■ *Amritsar (Punjab)*

SINGH, Gurnam (M/46)
Dept. of Political Science, Guru Nanak Dev University
Amritsar 143005, India.
Tel: 91-183-258848-0183(W)
Fax: 91-183-258820(W)

Specialisation : International Politics.

Areas of Current Research:
International Politics with Special Reference to South Asia; Ethnic Dimensions in International Politics.

Main Work at Hand:
Religion, Ethnicity & Nationalism in International Politics; Nationalism & Ethno-Cultural Diversity & Conflict.

Recent/Forthcoming Publications:
"Modernization, Ethnic Upsurge & Conflict in the World", in *International Journal of Group Tensions*, Vol.24, No.4, 1994.

■ *Azamgarh (UP)*

KHAN, Baber Ashfaq (M/30)
233, Baz Bahadur, Azamgarh 276001
India.

Specialisation : Defence Studies.

Areas of Current Research:
South Asian Strategic & International Studies; Indian Military History.

Main Work at Hand:
Generalship of Shahabuddin Mohammed Ghori; Nuclear Problem - India's Policy.

■ *Bangalore (Karnataka)*

CHENGAPPA, B.M. (M/35)
Principal Correspondent, The Hindu/Buisness Line
19 & 21, Infantry Road, Bangalore 560001, India.

Specialisation : Sino-Indian Relations.

Areas of Current Research:
Indian National Security & Foreign Policy; Indian Military Industry; Indian Defence Organization.

Recent/Forthcoming Publications:
"Indian Military Industry: Issues & Imperatives", in *Indian Defence Review*.

JAYARAMU, P.S. (M/42)
Professor, Dept. of Political Science
Bangalore University, Bangalore 560056, India.
Tel: 91-80-3355036 EXT.263(W) 3384677(H)
Fax: 91-80-3389295(W)

Specialisation : Indian Foreign Policy.

Areas of Current Research:
International Relations - Nuclear & Strategic Issues; South Asian Studies;
Indian Foreign Policy.

Main Work at Hand:
Indian Foreign Policy in Post Cold War Phase; Prospects for Peace in South
Asia.

Recent/Forthcoming Publications:
Article in Ranabir Samaddar (ed.), *Cannons into Ploughshares, Militarization
& Prospects for Peace in South Africa*, (New Delhi, Lancers Publications, 1995);
Article in Rama Melkote ed. *Indian Ocean Issues for Peace*, (New Delhi, Manohar,
1995).

KADAM, Umesh Veersen (M/39)
National Law School of India Post Bag No. 7201
Nagarbhavi, Bangalore 560072, India

Specialisation : International Space Law.

Areas of Current Research:
Space Law; Peace & Security Studies; Environmental Law.

Main Work at Hand:
Militarization of outer Space; Drafting National Space Legislation for India.

Recent/Forthcoming Publications:
"Espionage with Modern Technology & International Law", Vol. 7, *National
Law School* Journal 1995; "Protection of the Coastal Environment in Interna-
tional Law of the Sea", Vol.2, SBRRM *Journal of Law* 1995; "Liability for
Damage Caused by a Catastrophic Nuclear Accident", Ninon Franco (ed.),
Nuclear Energy & Sustainable Development, 1994.

RAMESH, M.K. (M/39)
Senior Assistant Professor,
National Law School of India University,
Nagaraghavi, Bangalore 560072, India.
E-mail: mkr@nls ernet.in

Specialisation : Environmental Law.

Areas of Current Research:
Environmental Law; Human Rights Law; International Law & International Relations.

Main Work at Hand:
Indigenous People & the Legal Order: Emerging Eco-Ethno-Ethics; Pollution Control Laws - A Comparative Study; Case Book on Private International Law.

Recent/Forthcoming Publications:
An Overview of Pollution Control Laws in India; Draft Bills on Land Aquisition, Resettlement & Rehabilitation; Environmental Law Reader.

SHASTRI, Sandeep (M/35)
Dept. of Social Science, Bangalore University
Bangalore 560056, India.

Specialisation : Indian Politics.

Areas of Current Research:
Comparative Politics in South Asia; Focusing on Federalism & Election Studies.

Main Work at Hand:
Emerging Trends in the Dynamics of Federation in South Asia; The Democratization Process in South Asia.

Recent/Forthcoming Publications:
Leadership & Dynamics of Federalism in South Asia; Electoral Politics in South Asia: Emerging Trends.

VINOD, M.J. (M)
Associate Professor, Dept. of Political Science
Bangalore University, Bangalore 560056, India.

Areas of Current Research:
US-India Relations; South Asian Security - Problems, Prospects; India: Foreign & Defence Policy.

Main Work at Hand:
US-India Relations: Challenges & Opportunities; Nuclear Proliferation in South Asia: Problems & Prospects.

Recent/Forthcoming Publications:
US Foreign Policy Towards India: A Diagnostic of the American Approach, (Lancers, New Delhi); *Nuclear Proliferation in South Asia: Current & Future Trends*, (University of Maryland).

■ Baroda (Gujarat)

ANADKAT, Nalin Karshandas (M/51)
Associate Professor, Department of Political Science
University of Baroda, India.

Specialisation : International Relations.

Areas of Current Research:
Strategic; Conflict; Peace Studies.

Recent/Forthcoming Publications:
Gujral Doctrine: India's Neighbours; Global Order & Peace: A Study of the
Perspectives of Ghandhi, Nehru & Lohia.

MOHITE, Dilip (M/55)
Dept. of Political Science, Faculty of Arts
M.S. University, Baroda 390002, India.
Tel: 329831(H)

Specialisation : International Relations.

Areas of Current Research:
International Relations Theory; Foreign Policy Analysis;
Regional Conflict (South Asia).

Main Work at Hand:
Prospects of Peace in South Asia.

Recent/Forthcoming Publications:
Prospects of Peace in South Asia; A Case of Conflict & Cooperation in Indo-
US Relations: Some Critical Issues 1994; India & the Emerging World Order
1995.

■ Berhampur (WB)

MAHANTY, Jinendra Nath (M/39)
Reader in Political Science, Berhampur University
Berhampur 760007, India.
Tel: 680-206924(W) 680-206924(H)

Specialisation : Chinese Foreign Policy.

Areas of Current Research:
International Politics; Chinese Foreign Policy; India's Foreign Policy/India &
South Asia.

Main Work at Hand:
Sino-Indian Relations: Post Cold War Period; Soviet Factor; Sino-Indian
Relations: The Cold War Years.

Recent/Forthcoming Publications:
"India & the Gulf Crisis", *Pacific Affairs*.

MAHIPATRA, Pradeep Chandra (M/40)
Lecturer in Journalism & Mass Communication, Berhampur University
P.O. 22, GPO, Berhampur 760001.
Tel: 91-680-203944(H)
Fax: 91-680-201959(H)

Specialisation : English Literature; Mass Communication.

Areas of Current Research:
Oriya Journalism; Journalism in South Asian Countries.

Main Work at Hand:
Development of Modern Oriyan Journalism.

PARICHA, Amiya Kumar (M/41)
Reader in Political Science, Berhampur University
Berhampur 760007, Orissa, India.
Tel: 91-680-209433(H)

Specialisation : Electoral Behavior.

Areas of Current Research:
Regional Cooperation in South Asia; India's Nuclear Policy; India's Relations with Her Neighbours.

Main Work at Hand:
Nuclear Weapons in International Politics, Policy Option for India; India's Regional Diplomacy in South Asia, SAARC.

SNEHALATA, Panda (F)
Dept. of Political Science, Berhampur University
Berhampur 760007, India.

Specialisation : Political Science.

Areas of Current Research:
International Relations; Political Pscychology; Women's Studies.

Main Work at Hand:
Nuclear Policy: India's Strategic Perceptions.

Recent/Forthcoming Publications: "US Foreign Policy Strategies in the Post Cold War Period", *Strategic Analysis,* Jan 1996.

■ *Bhagalapur (Bihar)*

KUMAR, Vijay (M/42)
Dept. of Political Science, Bhagalpur University
Bhagalpur 812007, India.
Tel: 91-641-423394(W) 91-641-423562(H)

Specialisation : International Politics.

Areas of Current Research:
Govt. & Politics in Sri Lanka; Regional Movements in South Asia;
Development Studies in South Asia.

Main Work at Hand:
Corruption in Politics: Indian Experiences; Impact of Liberalization &
Globalization on Indian Culture.

Recent/Forthcoming Publications:
Politics of Participation:; Development & Environment as Tools of International Politics; Politics of Secularism in India.

■ *Bhubaneswar (Orissa)*

HANS, Asha (F/50)
P.G. Dept. of Political Science, Utkal University
Vani Vihar, Bhubaneswar 751007, India.
Tel: 91-674-48021(W) 402186(H)
Fax: 91-674-480020(W) 405875(H)
E-mail: ahans@smrc.ori.nic.in

Specialisation : International Politics.

Areas of Current Research:
Refugee Studies.

Main Work at Hand:
Sri Lankan Refugees in India; Women Refugees; Protection of Rights in
Developing Countries.

Recent/Forthcoming Publications:
*Sri Lankan Refugees in India; Controling State Crime in India & Sri Lanka;
Responsibility Sharing: Study for Reformulation of Refugee Law*, (York
University).

KAR, Gyana Chandra (M/53)
Professor, Analytical & Applied Economic Dept.
Utkal University, Vani Vihar, Bhubaneswar 751004, India.
Tel: 91-674-601834(H)

Specialisation : Economics.

Areas of Current Research:
Development Economic & Rural Development.

Main Work at Hand:
Poverty & Human Development; Land Reforms.

MISRA, Surya Narayan (M/46)
Dept. of Political Science, Utkal University
Bhubaneswar 751004, India.

Specialisation : Politics.

Areas of Current Research:
Political Process; International Politics.

Main Work at Hand:
Party Politics & Electoral Choice in an Indian State; India: The Cold War
Years.

Recent/Forthcoming Publications:
Constitution & Constitutionalism in India (ed.); *China: Trends in Culture &
Development* (ed.).

NAYAK, Smita (F/31)
Lecturer, P.G. Dept. of Political Science
Utkal University, Bhubaneswar, India.

Specialisation : Political Science.

Areas of Current Research:
Project on National Election Studies 1996; Project on Democracy & Social
Capitalism in Segmented Societies.

Main Work at Hand:
Democracy & Social Capital in Segmented Societies; Tribal Studies.

PATNAIK, Sudhansu Kumar (M/45)
Reader, Dept. of Political Science
Utkal University, Bhubaneswar, India.

Specialisation : International Relations.

Areas of Current Research:
South Asia; South East Asia; International Relations Theory.

Main Work at Hand:
Encylopaedia Asiatic; Polisario Movement.

Recent/Forthcoming Publications:
APEC & USA in 21st Century; Polasario's Foreign Policy Behaviour.

RATH, Sharada (F/54)
Senior Professor, Dept. of Political Science
Utkal University, Bhubaneswar 751004, India.
Tel: 91-674 408544(H)

Specialisation : Political Science.

Areas of Current Research:
Comparative Govt. & Politics; Political Sociology; Women Studies.

Main Work at Hand:
Values of State Women Adminstrators in the US; Comparative Federalism.

Recent/Forthcoming Publications:
Research Methods in Social & Political Research; Women State Administrators in the US.

RAY, Rita (F/45)
Professor of Sociology, Utkal University
Vani-Vihar, Bhubaneswar 751004, India.
Tel: 91-674-482496(W) 481496(H)
Fax: 91-674-411020(W)

Specialisation : Sociology.

Areas of Current Research:
Social Anthropology; Gender Studies; Environmental Studies.

Recent/Forthcoming Publications:
Underground Drama: A Socio-Ecological Study of Two Chromite Mines in Orissa, (Delhi, Ajanta, 1944); *Socio-Economic Study of the New Components of the Social Forestry Project in Orissa* (Utkal University, 1966); "Impact Assessment of Women's Enterprise Management Training" (1996, World Bank Study).

SATAPATHY, Brahamananda (M/45)
Reader, P.G. Dept. of Political Science
Utkal University, Bhubaneswar 751004, India.

Specialisation : Non-Alignment.

Areas of Current Research:
International Politics; Women Studies; Democracy & Social Capital in Segmented Societies.

Main Work at Hand:
Non-Alignment: Dynamics & Future; India & CTBT.

Recent/Forthcoming Publications:
Non-Alignment: Dynamics & Future, National Publishing House, New Delhi.

■ *Calcutta (WB)*

BANERJI, Arun Kumar (M/51)
Dept. of International Relations, Jadavpur University
Calcutta 700032, India.
Tel: 91-33-473-4044(W)
Fax: 91-33-473-1484(W)

Specialisation : International Relations.

Areas of Current Research:
Strategic & International Relations in South Asia, including the Indian
Ocean; Politics in West Asia; US Foreign Policy.

Main Work at Hand:
India, Pakistan & the Peace Process.

Recent/Forthcoming Publications:
"The US & South Asia", S. Ghosh & S. Mukherjee (eds.), *Emerging South
Asian Order*, (Media South Asia, Calcutta, 1995); "From Camp David to
Cairo: The Peace Process in West Asia", *Jadavpur Journal of International
Relations,* Vol.1, 1995; "Images of Obsessions: India & Pakistan as Regional
Adviseries", S. Mansingh (ed) *Indian and Chinese Foreign Relations.*

BASU, Raj Sekhar (M/31)
Rabindra Bharati University, 56A Barrackpore Trunk Road
Calcutta 700050, India.

Specialisation : History.

Areas of Current Research:
Indo-US Studies; India & South East Asia; Ethnic & Minority Politics in
Bangladesh.

Main Work at Hand:
Minority Politics in Bangladesh.

Recent/Forthcoming Publications:
Modernisation & Resurgence of Insurgency in Chittagong Hill Tracts.

BASU, Gautam (M)
Professor, Dept. of International Relations
Jadavpur University, Calcutta 700032, India.

Specialisation : Political Science.

Areas of Current Research:
Theory Building in International Relations; Political Economy of Development in South Asia; Foreign Policy Making.

Main Work at Hand:
Alternative Theory Building in Third World Perspectives; How Foreign Policy Decisions are Made: Towards a Critical Perspective.

Recent/Forthcoming Publications:
The State, Development & Military Interventions; Theories of International Relations: Search for Alternatives; Pakistan: The Political Economy of Development.

BASU RAY CHAUDHURY, Sabyasachi (M/32)
Dept. of Political Science, Rabindra Bharati University
56A, B.T. Road, Calcutta 700050, India.
Tel: 91-33-3373732(H)
Fax: 91-33-2480178(W)

Specialisation : Political Economy.

Areas of Current Research:
Chittagong Hill Tracts Problem & it's Implications; US Policy in South Asia; SAARC & Regional Cooperation in South Asia.

Main Work at Hand:
Chittagong Hill Tracts Problem; US Policy in South Asia.

Recent/Forthcoming Publications:
Indo-US Economic Relations in an Era of Liberization; Caste in Indian Politics.

BHATTACHARYA, Purusottam (M/45)
Dept. of International Relations, Jadavpur University,
Calcutta 700032, India.
Tel: 91-33-473-4044(W), 440-5989(H)

Specialisation : West European Studies.

Areas of Current Research:
Western Europe; Indian Foreign Policy; South Asian Regional Politics.

Main Work at Hand:
European Union & South Asia; India & APEC.

Recent/Forthcoming Publications:
"India & Germany in International Studies", *New Delhi*, (Apr-Jun 1995).

BHATTACHARYA, Sanjukta Banerji (F/44)
Dept. of International Relations, Jadavpur University
Calcutta 700032, India.

Specialisation : International Relations.

Areas of Current Research:
Third World: American Studies; West Asia.

Main Work at Hand:
Deferred Hopes: Blacks in Contemporary America, 1986.

Recent/Forthcoming Publications:
Regional Security ,Threat Perceptions, Sub - Nationalism: India, Sri Lanka & the Tamil Ethnic Crisis, (School of International Relations & Strategic Studies, Jadavpur University, Calcutta); *American National Security Interests & Perception*.

BHATTACHARYA, Sapna (F/43)
Dept. of South & South-East Asian Studies, Calcutta University
51/1 Hazra Road, Calcutta 700019, India.
Tel: 91-33-4758895(W) 5502167(H)
Fax: 91-33-4408656(H)

Specialisation : History.

Areas of Current Research:
Myanmar: History & Contemporary Developments; Bangladesh: Ethnic Minorities.

Main Work at Hand:
Myanmar-India Relations; British Policy & the Tribal Population in India, Bangladesh & Myanmar.

Recent/Forthcoming Publications:
Bangladesh-Myanmar Relations: A Study of the Problem; "Refugees in Bangladesh", *Integration, Disintegration & World Order*, (By A. Benerjee, Allied Pub. in Collaboration with School of International Relations, Jadavpur University, Calcutta 1995); *Imperialist Provocation & Muslims of Arakan*, (Myanmar 1942-48).

BHATTACHARYYA, Rupak (M/27)
Senior Research Fellow,
Maulana Abul Kalam Azad Institute of Asian Studies,
567, Diamond Harbour Road, Behala, Calcutta 700034, India.
Tel: 91-33-4681396(W)

Areas of Current Research:
Political Role of the Bangladesh Armed Forces; Various Political & Social

Issues in Bangladesh; Security & Strategic Issues in South Asia.

Main Work at Hand:
The Military Rule in Bangladesh: A Study of it's Impact on Polity; Rohingya Refugee Problem in Bangladesh.

BOSE, Tarun Chandra (M/64)
26, Lansdowne Terrace, Calcutta 700026
India.
Tel: 91-33-4645074(W)

Specialisation : International Relations.

Areas of Current Research:
Defence/Strategic Studies; South Asia/ Environmental Security; Superpower Relations.

Main Work at Hand:
Environmental Security; Nuclear Weapons & Peaceful Uses of Atomic Energy; South Asian Dilemma.

Recent/Forthcoming Publications:
"Twenty Five Years of NPT: What Next ?", *Jadavpur Journal of International Relations,* Vol.1, 1995; "The Nuclear Isuue: US & Indian Perspective", Annpurna Nautiyal, ed., *India & The New World Order*, (New Delhi, South Asian Publishers, 1996).

CHAKRABARTI, Ranjan (M/36)
Dept. of History, Jadavpur University
Calcutta 700032, India.

Specialisation : History.

Areas of Current Research:
Crime & Society in Colonial Bengal; Early American Trade in Bengal.

Recent/Forthcoming Publications:
Authority and Violence in Colonial Bengal, 1800-1860; Political Economy & Protest in Colonial India.

CHAKRABARTY, Bhaskar (M/40)
Dept. of History, 51/2 Hazra Road
Calcutta 700019, India,

Specialisation : History.

Areas of Current Research:
Modern Indian History; Indian Nationalism; International Relations.

Main Work at Hand:
Migration & Political Stability in Eastern India; Intellectual History & India in the Twentieth Century.

Recent/Forthcoming Publications:
Fort William - A Historical Perspective.

CHAKRABORTI, Rajagopal Dhar (M/36)
Head, Dept. of South & Southeast Asian Studies
Calcutta University, 51/1 Hazra Road, Calcutta 700019, India.

Specialisation : Demography.

Areas of Current Research:
Population Ageing; International Migration; Population Policy.

Main Work at Hand:
Population Policy & Programmes of Thailand; Population Ageing in Asia.

Recent/Forthcoming Publications:
Incentive & Disincentives in Population Policies of Asia; Asian Mega Cities.

CHAKRABORTI, Tridib (M/37)
Dept. of International Relations, Jadavpur University
Calcutta 700032, India.
Tel: 91-33-4729766(H)
Fax: 91-33-4731484(W)

Specialisation : International Relations.

Areas of Current Research:
Strategic, Political, Economic & International Relations of South, Southeast Asia and Asia Pacific Region.

Main Work at Hand:
Constitutional Development of Vietnam: From Confrontation to Accommodation; Malaysia's Vietnam Policy Since 1965; Marginal Groups in a Plural Society: A Case Study of Indian Malaysians.

Recent/Forthcoming Publications:
"Three Stages in Indo-Vietnam Relations: An Overview", Santimoy Royand Tridit Chakraborthi (ed.); *Vietnam: A Changing Horizon*, (Chaterjee Publisher, Calcutta, 1992); "India's Southeast Asia Policy in the 1980's & 1990's: Contrasting Shades in Foreign Policy Priorities", *Jadavpur Journal of International Relations*, Vol.1, 1995, Dept. of International Relations, Jadavpur University, Calcutta.

CHAKRAVARTY, Papia (F/54)
Dept. of History, Jadavpur University
Calcutta 700032, India.

Areas of Current Research:
Social, Cultural, Political History of Bengal/India; Religious Interactions -
India & US, A Cross Cultural Study.

Main Work at Hand:
Quest for Rational Religion & Indian Nationalism, 1893 - 1939.

CHATTERJEE, Shibashis (M/25)
Dept. of Political Science, Rabindra Bharati University
56 A, B.T. Road, Calcutta 700050, India.
Tel: 91-33-4731109(H)

Specialisation : International Relations.

Areas of Current Research:
Indian Foreign Policy; International Relations Theory & International
Political Economy; South & South East Asia.

Main Work at Hand:
International Relations: From State to State; Multilateralism & Security
Complexes in South East Asia.

Recent/Forthcoming Publications:
The North Korean Nuclear Crisis; "Why Decision & Not Policy: Continuity
& Change in Indian Foreign Policy", *Rabindra Barati Journal*;
"Authoritaniasm in India", *Journal of Constitutional & Parliamentary
Studies*, New Delhi.

CHATTOPADHYAY, Anirban (M/37)
Ananda Bazar Patrika, 6, Prafulla Sarkar Street
Calcutta 700001, India.
Tel: 91-33-278000 Ext. 313(W), 778036(H)

Specialisation : Economics.

Areas of Current Research:
Economic Development & Social Empowerment, particularly in India;
Women & Work.

Main Work at Hand:
A Study on Economic Reform & Governance in South Asia; A Study on the
Impact of Development on Scheduled Cases & Tribes in West Bengal.

DAS, Suranjan (M/41)
Dept. of History, University of Calcutta
51/2 Hazra Road, Calcutta 700019, India.
Tel: 91-33-746730(W) 3591616(H)
Fax: 91-33-4407669(H)

Specialisation : History.

Areas of Current Research:
Nation Building & Regional Politics in South Asia; Confidence-Building
Measures in South Asia.

Main Work at Hand:
Ethnicity, Nation-Building & Regional Politics in South Asia: Kashmir,
Sindh & Bangladesh.

Recent/Forthcoming Publications:
"Kashmir Re-visited: Ethnicity Nation-Biulding & Regional Security in
South Asia" in *Ethnic Studies Report*, (Sri Lanka).

DAS, Rochana (F/32)
Dept. of International Relations, Jadavpur University
Calcutta 700032, India.
Tel: 91-33-4722170(H)

Areas of Current Research:
Ethnic Politics in South Asia; Government & Politics of India;
Separatist Problem in Northeast India.

Main Work at Hand:
Ethnic Politics in South Asia: India & Pakistan; Domestic Threat to Security:
A Case Study of Northeast India.

Recent/Forthcoming Publications:
"Ethnic Politics in South Asia: A Case Study of Assam & Sindh", *Jadavpur
Journal of International Relations,* Calcutta, India.

DASGUPTA, Subhendu (M/48)
Dept. of South & Southeast Asian Studies, University of Calcutta
Law Building, (2nd Floor), 51/1 Hazra Road, Calcutta 700019, India.
Tel: 91-33-4758895(W), 4731843(H)

Specialisation : Economics.

Areas of Current Research:
Migration in South Asia; Military in Bangladesh; Alternative Development in
South Asia.

Main Work at Hand:
Migration in South Asia; Ethnic Minorities in Bangladesh.

Recent/Forthcoming Publications:
"Making of an Alliance: Bourgeoisie & Army in Bangladesh", *The Journal of Social Studies,* No.71, Jan 1996; "Construction of the Insulation Theory: An Essay on Other Defence for Bangladesh", *Bulletin of the Department of the South & Southeast Asian Studies,* University of Calcutta, Vol.1, No.1, 1995.

DE, Amalendu (M/66)
Former Guru Nanak Professor of Indian History, Dept. of History
Jadavpur University, Calcutta, India.
Tel: 91-33-462-4249(H)

Specialisation : Indian History

Areas of Current Research:
Indian Freedom Struggle; Muslim Politics in India; Socio-Cultural Life in India; Indo-Bangladesh Relations.

Main Work at Hand:
Roots of Separatism in Nineteenth Century Bengal; Islam in Modern India.

Recent/Forthcoming Publications:
Religious Fundamentalism & Secularism in India; History of the Khaksar Movement in India; Swadeshi: In the Perception of the Bengali Muslims.

DE, Barun (M/63)
Maulana Abul Kalam Institute of Asian Studies
567, Diamond Harbour Road, Behala, Calcutta 700034, India.

Specialisation : Indian History.

Areas of Current Research:
Modern India from the 18th to the 20th Centuries; World History with Particular Reference to the Transition from Absolutism to Nationalism; Modern Central Asia from the South Asian Perspective.

Main Work at Hand: A History of Modern Central Asia from the South Asian Perspective.

Recent/Forthcoming Publications:
Socio-Political Character of Dual Governance in Early Modern India; Central & North West Asian Geopolitics in Post-USSR International Relations; Unity/Deversity: An Aspect of the Construction.

DUTTA, Nilanjan (M/35)
c/o Ananda Bazar Patrika Ltd, 6, Prafulla Sarkar Street
Calcutta 700001, India.
Tel: 91-33-27-8000(W) 91-33-440-9774(H)
Fax: 91-33-225-3240(W)
E-mail: nilanjan@ilcal.unv.ernet.in

Areas of Current Research:
Human Rights in South Asia; Communalism in South Asia; Water Diplomacy in South Asia.

Main Work at Hand:
History of Civil Liberties Movement in the Indian Subcontinent.

Recent/Forthcoming Publications:
From Subject to Citizen: Civil Liberties Movement in Colonial India.

GHOSH, Anjali (F/48)
Dept. of International Relations, Jadavpur University
Calcutta 700029, India.
Tel: 91-33-466-2142(H)

Specialisation : Political Science.

Areas of Current Research:
South Asia; Southeast Asia; US Policy in South & Southeast Asia.

Main Work at Hand:
State Capitalism in Indonesia; Regional Cooperation in South Asia.

Recent/Forthcoming Publications:
A Study of Maxist Politic in West Bengal, 1967-1977, Calcutta, 1981; Burma: A Case of Aborted Development, 1989; State Capitalism in Indonesia.

GHOSH, Lipi (F/37)
Senior Lecturer, Dept. of South & South-East Asian Studies
51/1 Hazra Road, Calcutta 700019, India.
Tel: 91-33-4758895(W), 3346469(H)

Specialisation : History.

Areas of Current Research:
Ethnicity & Nation Building in North-East Indian Politics.

Main Work at Hand:
Minorities at Mobilisation: Thai-Ahome of North-East India; Prostitution in Thailand: Perspective, Problem & Prevention.

Recent/Forthcoming Publications:
"Ethnicity, Religion & Identity Questions: A North East Indian Profile", in Journal of Political Science Department.

GHOSH, Nirmal Kanti (M/53)
Dept. of Political Science, Vidyasagar Evening College
Calcutta 700006, India.
Tel: 91-33-4729417(H)

Specialisation : Political Science.

Areas of Current Research:
South Asia; India & Her Neighbours.

Main Work at Hand:
Politics of Integration in South Asia.

Recent/Forthcoming Publications:
India: Foreign Policy Options.

GHOSH, Parimal (M/43)
Dept. of South & Southeast Asian Studie, Calcutta University
51/1, Hazra Road, Calcutta 700019, India.
Tel: 91-33-4758895(W) 4733244(H)

Specialisation : Labour History.

Areas of Current Research:
History of Industrial Labour, Bengal; Peasant Rebellions, Burma (Colonial Period).

Main Work at Hand:
Class, Community & Colonial Labour, Bengal, 1880-1980; Peasant Revolts in Colonial Burma.

GHOSH, Sucheta (F/46)
Dept. of International Relations, Jadavpur University
Calcutta 700032, India.
Tel: 91-33-4758136(H)

Specialisation : International Relations.

Areas of Current Research:
South Asia.

Main Work at Hand:
Developmental Regionalism; Indian Foreign Policy.

MAJUMDAR, Madhumita (F/28)
Maulana Abul Kalam Azad Institute of Asian Studies
567, Diamond Harbour Road, Calcutta 700034, India.
Tel: 91-33-72-3985(H)

Specialisation : History.

Areas of Current Research:
History; Sociology; Cultural Studies.

Main Work at Hand:
Sciences & the Nationalist Agenda in Bengal 1935-47; The Making of India's
National Cultural Institutions 1944-54.

MISHRA, Omprakash (M)
Dept. of International Relations, Jadavpur University
Calcutta 700032, India.

Specialisation : International Politics.

Areas of Current Research:
Regionalism & Regional Integration; United Nations; Refugees.

Main Work at Hand:
Political Economy of Regional Integration in the Third World;
Developmental Regionalism in South Asia.

Recent/Forthcoming Publications:
"Political Economy of Asia Pacific Cooperation", *Jadavpur Journal of
International Relations,* Vol.1, 1995; Refugee Problem & the UNHCR;
SAARC: Crisis & Survival.

MUKHERJEE, Rila (F/39)
Dept. of History, Jadavpur University
Calcutta 700032, India.

Specialisation : Economic Sciences.

Areas of Current Research:
State Formation & Culture; Reconstruction of Criminality.

Main Work at Hand:
Court & Polity.

Recent/Forthcoming Publications:
Indian Historical Review, Vol.19 & 20; Contributions to *Indian Sociology*
1996 & 1997.

MUKHERJEE, Somen (M/55)
Flat No.232, 3, Bidhan Sishu Sarani
Calcutta 700054, India.

Specialisation : Philosophy.

Areas of Current Research:
Psychology of Regional Cooperation; Indian Philosophy with Reference to
Western Psychology.

Recent/Forthcoming Publications:
Emerging South Asian Order: Hopes & Concerns (ed.), (Calcutta, 1995).

RAGHAHARI, Chattergji (M/52)
Dept. of Political Science, Calcutta University
1, Reformatony Street, Calcutta 700027, India.

Specialisation : Political Science.

Areas of Current Research:
Working Class Politics in India; Ethnicity & International Relations in South Asia; Nation Building & State Formation.

Main Work at Hand:
Ethnicity & International Relations in South Asia.

Recent/Forthcoming Publications:
Religion, Politics & Communalism - The South Asia Experience (ed.), (New Delhi, South Asian Publishers, 1994).

RAI CHOWDHURI, Satyabarata (M/58)
Dept. of Political Science, Rabindra Bharati University
56, B.T. Road, Calcutta 700050, India.
Tel: 91-33-3371447(H)

Specialisation : Political Science.

Areas of Current Research:
International Relations, South Asia in Particular.

Main Work at Hand:
Leftist Movements in India; Post Cold War World.

RAY, Jayanta Kumar (M/61)
Dept. of History, Calcutta University
51/2 Hazra Road, Calcutta 700019, India.
Tel: 91-33-746730(W) 4733187(H)
Fax: 91-33-4739175(H)

Specialisation : Arts & International Relations

Areas of Current Research:
Public & International Affairs & Development Studies; Policy Studies.

Main Work at Hand:
India-Nepal Cooperation Broadening Measures; India-Bangladesh Cooperation Broadening Measures.

Recent/Forthcoming Publications:
Inside Bureaucracy: Bangladesh, (Co-author); *An Uncertain Beginning: Perspective on Parliamentary Democracy in Bangladesh*, (Calcutta, 1987); *Democracy in Bangladesh*, (Calcutta, 1992); *Civil Society in Bangladesh: Resilience & Retreat*, (Calcutta 1996).

ROY, Ranjit Kumar (M/38)
Dept. of History, Ravindra Bharati University
56-A, B.T. Road, Calcutta 700050, India.
Tel: 91-33-2429777(H)
Fax: 91-33-5568079(W) 91-33-2427282(H)

Specialisation : History.

Areas of Current Research:
Development Studies; Women's Issues; Peace & Conflict Resolution -
Application of Ghandian Principles.

Recent/Forthcoming Publications:
Bengal - Yesterday & Today (editor) 1990; *Imperial Emulous-Society &
Polity Under the Raj*, (editor) 1992; *Retrieving Bengali's Past - Society and
Culture in the 19th & 20th Centuries*, 1994. Non Indian Perception of
Ghandhi; *Towards a New Interpretation of Ghandhi* (ed.); *Civil Disobedience
& Student Politics in Bengal, 1920-47.*

ROY, Asish K. (M/51)
Dept. of South & Southeast Asian Studies, Calcutta University
Law College Building, 51/1 Hazra R, Calcutta 700019, India.

Specialisation : Communist Movement in India.

Areas of Current Research:
Political Development, Military Politics & Narcotics Problems in South &
Southeast Asia.

SAMADDAR, Ranabir (M/46)
Maulana Abul Kalam Azad Institute of Asian Studies
567, Diamond Harbour Road, Calcutta 700034, India.
Fax: 91-33-468-1396(W)

Specialisation : Political Science.

Areas of Current Research:
New Technology & Labour; Nationalism in South Asia, particularly Bangladesh; Peace Laws.

Main Work at Hand:Migration Studies - Transborder Migration from
Bangladesh to West Bengal.

SARKAR, Mahua (F/37)
Dept. of History, Jadavpur University
Calcutta 700032, India.
Tel: 91-33-4656278(H)

Specialisation : History of Bengal.

Areas of Current Research:
Judicial & Legal History of South Asia; Environmental Policies & Eco-Feminism; South Asian Life & Thought in 20th Century.

Main Work at Hand:
Deforestation & Eco Feminism in India; Life & Thought in 20th Century Bengal.

Recent/Forthcoming Publications:
Justice in Bengal, 1861-1915; Deforestation in South Asia.

SENGUPTA, Anita (F/26)
Maulana Abul Kalam Azad Institute of Asian Studies
567, Diamond Harbour Road, Behala, Calcutta 700034, India.
Tel: 91-33-3372677(H)
Fax: 91-33-4681396(W)

Specialisation : Political Studies

Areas of Current Research:
Central Asian History, Politics & Culture, particularly of the 19th & 20th Centuries.

Main Work at Hand:
Language, Religion & State Formation in Central Asia - A Case Study of Uzbekistan.

Recent/Forthcoming Publications:
"After the Heartland & it's Southern Rim: Central & South Asia Today", in Ranabir Samaddar (ed.), *Cannon Into Ploughshares, Militarization & Prospects of Peace in South Asia*; "The Unchanging Face of Central Asia", in *Maulana Abdul Kalam Azad Institute of Asian Studies Newsletter* Vol.1, 1994; Russians as Minorities in Central Asia.

TAJUDDIN, Mohammad (M/33)
Maulana Abul Kalam Azad Institute of Asian Studies
567, D.H. Road, Calcutta 700034, India.

Specialisation : Diplomacy ; International Relations.

Areas of Current Research:
Bangladesh Foreign Policy & Foreign Relations; Security Policies of the Smaller States of South Asia; Domestic Politics of Bangladesh.

Main Work at Hand:
Socio Political Study of the Ethnic Minorities in Bangladesh.

Recent/Forthcoming Publications:
South Asia & China: An Overview & the Case Study of Bangla-China Relations; Secular Ends & Religious Means: Bangladesh & the Islamic

World; China's Third World Policy from Mao to Deng.

■ Darjeeling (WB)

CHAKRABARTY, Manas (M/45)
Dept. of Political Science, North Bengal University
Darjeeling, West Bengal, India.
Tel: 91-353-450471(W) 91-353-450322(H)
Fax: 91-353-450546(W)

Specialisation : Political Science.

Areas of Current Research:
International Relations; Comparative Judicial Behaviour; Women & Politics.

Main Work at Hand:
Women & Politics: A Study of Women Parliamentarians in India;
Judicial Behaviour & Judicial Decision Making of the Indian Supreme Court.

Recent/Forthcoming Publications:
"Women & Politics", *Punjab Journal of Politics*; "Caste & Politics, in Social
Change; Elections & Voting Behaviour", *Indian Journal of Political Science*.

■ Dharwad (Karnataka)

PATAGUNDI, Shivaputra (M/42)
Reader in Political Science, Karnataka University
Dharwad 580003, Karnataka, India.
Tel: 91-836-347121 Ext.23(W)
Fax: 91-8-836-348464(W)

Specialisation : Political Science.

Areas of Current Research:
International Relations; Foreign Policy Studies; Party System & Political Process.

Recent/Forthcoming Publications:
Political Parties, Party System & Foreign Policy of India, (New Delhi: Deep
& Deep, 1994); *India's Foreign Policy: An Elitist Perception*, New Delhi:
(Uppal Publishing House, 1994). *Outlines of Modern Political Analysis*,
(New Delhi, Oxford & IBH Publishing Co.); Indian Secularism: Crisis in
Theory & Practice.

RAMASWAMY, Harish (M/35)
Dept. of Political Science, Karnataka University
Pavate Nagar, Dharwad 580003, India.
Tel: 91-836-347121 Ext.23(W) 46422(H)
Fax: 91-836-347884(W)

Specialisation : Public Administration.

Areas of Current Research:
Local Government & Rural Development Including Decentralisation; Globalisation & Allied Aspects Like WTO, Bio-Diversity etc., in Indian Context; Geo-Political Issues Covering Foreign Policy & International Relations.

Main Work at Hand:
Emerging Trends in Rural Local Leadership in Response to Decentralised Administration in the State of Karnataka; National Election Study 1996.

Recent/Forthcoming Publications:
Distance Education; Fertility & Planning for Rural Development; Geo-Political Implications of the Convention on Bio-Diversity: The Indian Context.

■ *Garwhal (UP)*

NAUTIYAL, Annpurna (F/38)
Dept. of Political Science, P.O. Box 16
H.N.B. Garwhal University, Srinagar, Garwhal 246174, India.

Specialisation : Political Science.

Areas of Current Research:
Indo-US Relations; US & South Asia; Regional Political Issues.

Main Work at Hand:
End of Cold War & South Asia: A Study of Triangular Relationship Between India, Pakistan & US.

Recent/Forthcoming Publications:
"Kashmir Issue: The International Perspective", *Asian Studies* July-Dec 1995, Vol 12, No.2 Calcutta; *India & the New World Order*, 1995, (South Asian Publishers, New Delhi; *Uttarakhand: The Recent Turmoils*, (M.D. Publishers, New Delhi).

■ *Gauhati (Assam)*

BEZBARUAH, Ranju (M/48)
Professor & Head, Dept. of History, Gauhati University
Gauhati 781014, India.
Tel: 91-361-571479(H)

Specialisation : Indo-US Relations.

Areas of Current Research:
International Relations; American History; North-East Indian History.

Main Work at Hand:
Isolation to Global Community.

Recent/Forthcoming Publications:
Convergence & Divergence in Indo-US Relations.

GOGOI, Phani Dhar (M/59)
Dept. of Political Science, Gauhati University
Gauhati, India.

Areas of Current Research:
Local Institution; Ethnic Movement.

Main Work at Hand:
Study on Ethnicity in North-East India.

HUSSAIN, Monirul (M/44)
Reader in Sociology
Dept. of Political Science, Gauhati University
Gauhati 781014, India.
Tel: 91-361-570443(W) 91-361-571808(H)
E-mail: mhussain@atgu.ernet.in

Specialisation : Sociology.

Areas of Current Research:
National, Ethnic & Communal Question in South-Asia.

Main Work at Hand:
Religious Minorities of South Asia; Ethnic Conflict & Refugee Problem.

Recent/Forthcoming Publications:
The Assam Movement: Class, Ideology & Identity-1993; "Refugees in the Face of Emerging Ethnicity in North-East India", *Journal of Humanities & Social Sciences,* Vol.2, No.1, 1995; Roots of Ethnic Conflict & Violence in India's North-East.

MEDHI, Kunja (F/54)
Professor & Head, Dept. of Political Science
Gauhati University, Gauhati, India.

Specialisation : Political Science.

Areas of Current Research:
Legislative and Women's Studies; Human Rights.

Recent/Forthcoming Publications:
State Politics in India.

■ *Hyderabad (AP)*

ADLURU, Subramanyam Raju (M/34)
Research Scholar, Dept. of Political Science, University of Hyderabad
Andhra Pradesh, Hyderabad 500046, India.

Specialisation : Political Science.

Areas of Current Research:
International Relations; Indian Foreign Policy; Indian Ocean.

Main Work at Hand:
The US Policy Towards India 1969-1988; The French Presence in the
Western Indian Ocean Region in 1990.

Recent/Forthcoming Publications:
"France & the Indian Ocean: Zone of Peace", *Third Concept,* New Delhi,
Aug 1993; "The Indian Ocean Commission & France", *Journal of Indian
Ocean Studies,* New Delhi, March 1996; *The French Political Presence in
the Western Indian Ocean Region 1975-1990,* (1996).

AVASARALA, Neeraja (F/23)
Plot No. 18, Flat 2, Lake View Apartments, Prakashnagar
Begumpet, Hyderabad 500016, India.
Tel: 91-40-815820(W)

Specialisation : Political Science.

Areas of Current Research:
Indo-US Relations.

Main Work at Hand:
Presidential Elections in US: A Study of the Changing Trends.

BASAVAIAH, M. Channa (M/34)
Lecturer, Centre for Area Studies
Osmania University, Hyderabad 500007, India.
Tel: 91-40-7018951 Ext: 328(W)

Areas of Current Research:
International Relations, Indian Ocean Studies; South Asian Studies; African
Studies.

Recent/Forthcoming Publications:
"Civil Life in Ruwanda & Burundi", *Strategic Analyses,* Dec 1996;
"Thinking Along the Indian Coast", in *Indian Ocean: Issues for Peace,*
(Manohar Publication, New Delhi 1995).

GEORGE, Sudhir Jacob (M/52)
Dept. of Political Science, University of Hyderabad
Central University P.O., Hyderabad 500046, India.
Tel: 91-40-258501 EXT 3200(W)
E-mail: sjgss@unhyd.ernet.in

Specialisation : International Politics.

Areas of Current Research:
International Relations; South Asia; Foreign Policies; North East India.

Main Work at Hand:
Intervention in International Relations: Theory & Practice;
Unrest in North East India.

Recent/Forthcoming Publications:
The Chakma Refugees in Tripura, (New Delhi, South Asian Publishers, 1993); "The Bodo Movement in Assam: Unrest to Accord", *Asian Survey*, Oct 1994.

HARSHE, Rajendra Govind (M/46)
Dept. of Political Science, University of Hyderabad
Hyderabad 500046, India.
Tel: 91-40-258500 Ext 3200(W) 259551(H)
Fax: 91-40-2050(W)

Specialisation : International Relations.

Areas of Current Research:
Theories of International Relations; Politics in South Asia;
Sub-Saharan Africa; Western Europe.

Main Work at Hand:
Pervasive Entente, France & Ivory Coast in African Affairs.

Recent/Forthcoming Publications:
"West European Integration & the World Order", in Kanta Ahiya (ed.);
Strength for Mastering over Southern Africa, New Delhi Sage 1995.

MELKOTE, Rama S. (F/55)
Centre for Area Studies, Press Road
Osmania University, Hyderabad 500007, India.

Specialisation : International Relations.

Areas of Current Research:
International Relations; Political Theory; Area Studies - Indian Ocean.

Main Work at Hand:
Reconceptualising Peace - Issues for Indian Ocean; Meaning of Globalisation.

Recent/Forthcoming Publications:
Indian Ocean - Issues for Peace, (Manohar Publications, 1995); *The Third World City: Emerging Contours*, (Delta Publishers); *Political Economy & Foriegn Policy, A Study of Malagassy Foriegn Policy.*

RAO, Janagam Laxmi Narasimha (M/35)
Dept. of Political Science, Osmania University
Hyderabad 500007, India.

Specialisation : International Relations.

Areas of Current Research:
International Relations; Strategic Studies; Public Policy.

Recent/Forthcoming Publications:
Liberalization Syndrome: Uruguay Round & India.

RAO, P. Venkateshwar (M/45)
Professor of Political Science, Centre for Area Studies
Osmania University, Hyderabad 500007, India.

Specialisation : International Relations.

Areas of Current Research:
Indian Ocean Studies; South Asia; European Studies.

Main Work at Hand:
Indo-US Relations: Compessional Perspective; Reconceptualisation of Peace & Security in Indian Ocean; Regional Cooperation in South Asia: Role of Private Sector.

Recent/Forthcoming Publications:
Peace & Security in Indian Ocean.

REDDY, Gavva Ram (M/35)
Lecturer, Dept. of Political Science
O.U. College for Women, Koti, Hyderabad 500195, India.

Specialisation : Political Science.

Areas of Current Research:
International Relations; International Security Studies; International Law.

Main Work at Hand:
Nation-State of New International Order.

Recent/Forthcoming Publications:
South Asian Security - International Law.

SUSMITHA, (F/25)
Research Scholar, University of Hyderabad
Hyderabad 500046, India.

Specialisation : Political Science.

Areas of Current Research:
Regional Economic Organization; SAARC; ASEAN.

Main Work at Hand:
Regional Economic Organization: A Study of ASEAN.

VARANASI, Yoga Jyotsna (F/37)
Reader, Dept. of Political Science, Osmania University
O.S. College for Women, Koti, Hyderabad 500195, India.
Tel: 91-40-557813(W) 7616538(H)

Specialisation : Political Science.

Areas of Current Research:
International Relations; Strategic Studies; Area Studies.

Main Work at Hand:
Western & Non-Western Perceptions of Security: Implications For Indian
Policy Formulation.

■ *Jaipur (Rajasthan)*

BUDANIA, Rajpal (M/31)
South Asia Study Centre, University of Rajasthan
Jaipur 302004, India.
Tel: 91-141-511980(W) 91-141-510134(H)

Specialisation : India's Pakistan Policy.

Areas of Current Research:
Strategic & Conflict Resolution Studies; Indo-Pakistan Relations; India's
National Security & Foreign Policy.

Main Work at Hand:
India's Pakistan Policy: A Study in the Context of Security; Change in the
Nature of International Security: It's Impact on India.

Recent/Forthcoming Publications:
"The Post Cold War Era: Challenges to India's External Security", *South
Asian Studies*, Jan-Dec 1993; U.S. & South Asia: Strategic Concerns &
Interaction Patterns; Indo-Pakistan Relations: Alternate Approaches to Peace
& Conflict Resolution.

GHILDIAL, Nalineet (F/28)
UGC Research Fellow, South Asian Studies Centre
University of Rajasthan, Jaipur 302004, India.

Specialisation : Political Science.

Areas of Current Research:
International Relations; Role of Major Powers in South Asia; India & South
Asia, Conflict Resolution in South Asia.

Main Work at Hand:
In the Absence of a Balancing Power: Indo-Soviet Relations & their Impact
on South Asia; Indo Russian Relations in the Post Cold War Era.

Recent/Forthcoming Publications:
"Indo-Nepal Relations & China", in Ramakant, B.C. Upreti (eds.), *Indo-
Nepal Relations*, (New Delhi, South Asian Publishers, 1992); "Disintegration
of the Soviet Union: Implications for India", in Ramakant, P.L. Bhola (eds.),
Post Cold War Developments in South Asia, (Jaipur, RBA Publisher, 1951);
Pakistan-Russian Relations in the Post Cold War Era.

JAIN, B.M. (M/49)
Research Scientist (UGC), South Asia Studies Centre
University of Rajasthan, Jaipur 302004, Indian.
Tel: 91-141-652227(H)
Fax: 91-141-652784(H)

Specialisation : Indo-U.S. Relations.

Areas of Current Research:
International Relations; South Asian Defence, Security, Strategic & Foreign
Affairs; China's Security, Defence & Foreign Affairs.

Main Work at Hand:
South Asian Security in the New World Order; India's Defence & Security:
State, Society & Science.

Recent/Forthcoming Publications:
Nuclear Politics in South Asia: In Search of an Alternative Paradigm; South
Asia, India & the U.S.; South Asian Security: Problems & Prospects.

KAYATHWAL, Mukesh (M/30)
5-CH-12, Jawahar Nagar
Jaipur 302004, India.

Specialisation : Pak-Afghan Relations.

Areas of Current Research:
Information Technology in South Asia; Regional Security & Cooperation;
Foreign Policies in South Asia.

Recent/Forthcoming Publications:
Security & Foreign Policies in South Asia (ed.); New Developments in South Asia (ed.); UN: Retrospect & Prospects (ed.). Foreign Policy of Pakistan.

■ Madras (TN)

ANURADHA, C.S. (F/24)
Dept. of Defence & Strategic Studies, University of Madras, Chepauk, Madras 600005, India.

Specialisation : International Relations.

Areas of Current Research:
Non-Proliferation; South Asian Security; Indo-U.S. Relations.

Main Work at Hand:
NPT - A Factor in Indo-U.S. Relations.

BHASKARAN, S. (M/57)
Professor of Political Science, Annamalai University
Annamalai Nagar 608002, Tamil Nadu, India

Specialisation : Political Science.

Areas of Current Research:
Research Methodology; Comparative Governments; International Policy.

Main Work at Hand:
Politics in Poly-ethnic Polities; Disarmament & Development.

PRABHAKAR, W. Lawrence S. (M/31)
Assistant Professor, Dept. of Political Science
Madras Christian College, Tambaram, Madras 600059, India.
Tel: 91-44-2375675(W) 91-44-6280876(H)

Specialisation : International Politics; Strategic Studies.

Areas of Current Research:
International Politics & Foreign Policy; Strategic Studies; Ethnicity & Federalism.

Main Work at Hand:
Strategic Dynamics in India Since 1971; Ballistic Missile Proliferation - A Comparative Study of Middle East & South Asia.

Recent/Forthcoming Publications:
The Indo-Pak Ballistic Missile Race, *Asian Studies,* Jan-March 1991.

RAMASESHAN, Geeta (F/38)
III A, Chambers High Court
Madras - 600104, Tamil Nadu, India.
Tel:5341073(W) 4991397(H)
Fax:91-44-499039(H)

Specialisation : Law.

Areas of Current Research:
Human Rights & The Law; Family Law; Police & the Law; Media Rights.

Main Work at Hand:
Court Room Procedures on Rape Trials.

SURYANARAYAN, Venkateswaran (M/57)
Director, Centre for Southeast Asian Studies
University of Madras, Madras 600005, India.
Tel: 91-44-568778 Ext.323(W) 4916238(H)
Fax: 91-44-566693(W)

Specialisation : Southeast Asian Studies.

Areas of Current Research:
International Relations of South & Southeast Asia; Ethnicity & Nation
Building; Indian Minorities in South & Southeast Asia.

Main Work at Hand:
Maldives - Challenges of Development; The Other Tamils - Indian Tamils of
Sri Lanka.

Recent/Forthcoming Publications:
Kachchativu & Problems of Indian Fishermen in the Palk Bay Region, (T.R.
Publishers, Madras, 1995); "Chandrika Proposals: Rebuilding the Founda-
tions of a Plural Society", *Journal of Indian Ocean Studies,* Nov.1995; "Sri
Lankan Tamil Refugees in Tamil Nadu", S.D. Muni & Lok Raj Baral (eds.),
Refugees & Internal Security in South Asia, (Konark Publishers, New Delhi,
1996).

■ *Madurai (TN)*

JOHNSON, Albert (M/59)
Dept. of Political Science, Madurai Kamaraj University
Madurai 625021, India.
Fax: 91-452-85239(W)

Specialisation : Political Science.

Areas of Current Research:
Foreign Policy; Federalism; Constitutional Law.

91

Main Work at Hand:
Issues in Foreign Policy: A South Asian Perspective.

Recent/Forthcoming Publications:
Working of Federalism in India, HSRC, Pretoria, South Africa.

MADHANAGOPAL, R. (M/38)

Reader (Associate Professor), Dept. of Political Science
Madurai Kamaraj University Madurai 625021, India
Tel: 91-452-85407(W) 452-46242(H)
Fax: 91-452-85239(W)

Specialisation : International Relations.

Areas of Current Research:
South Asian Security; Nuclear Proliferation; Indo-American Relations;
Conflict Management & Resolution.

Main Work at Hand:
UN & the Use of Force: A Comparative Analysis of Korean & Gulf Crisis;
Dynamics of Proliferation & Prospects for Arms Control in South Asia.

Recent/Forthcoming Publications:
US Response to India's Nuclear Policy, 1963-1983, (Lancers Publishers).

■ Mhow (MP)

NAMBIAR, E.K.G. (M/48)

Dept. of History, University of Calicut
Malapuram Dt., Kerala, India.

Specialisation : International Relations.

Areas of Current Research:
Indian Foreign Policy; Russian Foreign Policy; International Relations.

Main Work at Hand: .
India and Nuclear Disarmament - Recent Trends; Indo-U.S. Relations.

Recent/Forthcoming Publications:
Role of Individuals in Re-Formulation of Indian Foreign Policy; 50 Years of
UN: An Evalution.

KHAN, Fazlur Rahman (M/48)

Professor of Military Science, Govt. Post Graduate College
Mhow 453441, India.
Tel: 91-7324-35325(H)

Specialisation : Military Geography.

Areas of Current Research:
Area Study - Indian Ocean; Disarmament, Arms Control & Peace; National Security.

Main Work at Hand:
National Security; Remote Sensing.

Recent/Forthcoming Publications:
Security of the Sub-Continent, Challenges from the Super Power.

■ *Mumbai (Maharashtra)*

ANAND, Javed (M/45)
Nirant Juhu Tara Road, Mumbai 400049
India.
Tel: 91-22-6053927(W)
Fax: 91-22-6482288(W)
E-mail: admin@sabrang.ilbom.ernet.in

Specialisation : Engineering.

Areas of Current Research:
Religious Fundamentalism; Progressive Muslim World; Human Rights.

Main Work at Hand:
Police & Communication.

SAVARA, Mira (F/47)
B-10 Sun-N-Sea, 25 JP Road, JP Road
Andheri, Mumbai 400026, India.

Specialisation : Sociology.

Areas of Current Research:
Public Health; Women; Environment.

Main Work at Hand:
Women, Organization & the Informal Sector

Recent/Forthcoming Publications:
Sexual Behaviour Patterns in India & AIDS Policy.

SETALVAD, Teesta (F/33)
"Nirant", Juhu Tara Road,
Mumbai 400049,
India.
Tel: 91-22-6053927(W)
Fax: 91-22-6482288(W)

E-mail: admin@sabrang.ilbom.ernet.in

Specialisation : Philosophy.

Areas of Current Research:
Religious Fundamentalism; Human Rights; Children & Prejudice;
Resolution of Conflicts.

Main Work at Hand:
The Hindu Women; Partition of the Subcontinent: A Human Perspective.

Recent/Forthcoming Publications:
Communication Combat.

THAKKAR, Usha (F/54)
Professor & Head, P.G. Dept. of Political Sciences,
SNDT Women's University
1, Nathibai Thackersey Road, Mumbai 400020, India.
Tel: 91-22-3625954(H)

Specialisation : Political Science.

Areas of Current Research:
Women & Politics; International Politics; Indian Politics.

Main Work at Hand:
Research Papers on Gandhi; Women in Indian Parliament.

Recent/Forthcoming Publications:
Politics in Maharashtra (Himalaya Pub. House, Mumbai, 1995); *Arab-Israel Accord* (Parichay Trust, Mumbai, 1995); *Fifty Years of U.N.*, (Parichay Trust, Mumbai, 1995).

WASLEKAR, Sundeep Veerkant (M/36)
I.C.P.I., B-704 Montana
Lokhandwala Complex, Andheri West, Mumbai 400053, India.
Tel: 91-22-6265672(W)
Fax: 91-22-6318260(W)

Specialisation : Commerce.

Areas of Current Research:
Governance & Democracy; Conflict Resolution; Economic Cooperation.

Main Work at Hand:
Vision 2022; Conflict Resolution Between India & Pakistan.

Recent/Forthcoming Publications:
A Handbook for Conflict Resolution in South Asia; South Asian Drama: Travails of Misgovernance, (Konark, 1995); *Track Two Diplomacy in South Asia*, (ACDIS, University of Illinois, 1995).

■ *New Delhi*

AGRWAAL, Ashok (M/37)
56, Todar Mal Road, New Delhi 110001
India.

Specialisation : Law

Areas of Current Research:
History, particularly, Civilisational Aspects; Ecology.

AHMAD, Imtiaz (M/54)
Centre for Political Studies, Jawaharlal Nehru University
New Delhi 110067, India.

Specialisation : Sociology.

Areas of Current Research:
Ethnic Conflict, Communalism; Islam, Muslim Social Institutions;
Education; Poverty.

Main Work at Hand:
The Uniform Civil Code Issue in India; Islamic Ideologies & Social Realities.

Recent/Forthcoming Publications:
Islamic Ideologies & Social Realities, (Delhi, Manohar).

ARORA, Balveer (M/50)
Centre for Political Studies, School of Social Sciences
Jawaharlal Nehru University, New Delhi 110067, India.
Tel: 91-11-670810(H)

Specialisation : Political Science.

Areas of Current Research:
Federation & Intergovernment Relations; Political Institutions;
Multistate & Regional Cooperation.

Main Work at Hand:
Bureaucracy & Federalism in India; Conflict & Cohesion in Transitional
Federal Systems.

Recent/Forthcoming Publications:
*Multiple Identities in a Single State: Indian Federation in Comparative
Perspective*, (Konark New Delhi 1995); *Federalism in India: Origins &
Development*, (Vikas, New Delhi 1992).

BARUA, Poonam Sethi (F/39)
Public Affairs Management, D-3/3492 Vasant Kunj
New Delhi 110070, India.
Tel: 91-11-6899930(W)

Specialisation : Economics.

Areas of Current Research:
Economic Confidence Building Between India & Pakistan; Regional Trading
Blocs & SAARC; Corporate Diplomacy.

Main Work at Hand:
Towards a Free Trade Arrangement in South-Asia.

Recent/Forthcoming Publications:
*Economic CBM's Between India & Pakistan in Crisis Prevention, Confidence
Building & Reconciliation in South Asia,* (Henry L. Stimson Centre, St.
Martin's Press, New York, November, 1995); *Towards a Free Trade Arrange-
ment in South Asia,* (Indian Institute of Foreign Trade, New Delhi, October
1995).

BEDI, Rahul (M/43)
A 14, Niti Bagh
New Delhi 110049, India.

Specialisation : English Literature.

Areas of Current Research:
Defence & Security Related Areas.

BEHERA, Ajay Darshan (M/31)
Assistant Research Professor, Centre for Policy Research
Dharma Marg, Chanakyapari, New Delhi 110001, India.
Tel: 91-11-3015273(W) 91-124-360742(H)
Fax: 91-11-3012746(W)

Specialisation : International Relations.

Areas of Current Research:
Conflict; Security; Peace.

Main Work at Hand:
Manufacturing Resolution: The JVP in Sri Lankan Politics.

Recent/Forthcoming Publications:
"An Analysis of Separatist Insurgencies in India", *Strategic Analysis,* January
1996; Politics of Violence and Development in South Asia.

BERRY, Varsha Rajan (F/26)
355 Ganga Hostel, Jawaharlal Nehru University
New Delhi 110067, India.

Specialisation : International Relations.

Areas of Current Research:
Women in Combat Services; Women & War: A Study of American Women
Since World War II.

BILGRAMI, S.J.R. (M/56)
Professor & Head, Dept. of Political Science
Jamia Millia, New Delhi 110025, India.
Tel: 91-11-6831717(W) 2433607(H)

Specialisation : India's Role in U.N.

Areas of Current Research:
International Politics; International Organization; South Asia.

Main Work at Hand:
Arms Race & Disarmament; Nehru, Indian Policy & World Affairs.

Recent/Forthcoming Publications:
United Nations in Encyclopaedia of Cold War, (Garner Publications); India,
UN & New World Order; United Nations with & without Cold War.

CHABHA-BEHERA, Navnita (F/28)
Assistant Research Professor .
Centre for Policy Research, Dharma Marg
Chanakyapuri, New Delhi 110021, India
Tel: 91-11-3015273(W) 91-124-360742(H)
Fax: 91-11-3012746(W)

Specialisation : Interantional Relations.

Areas of Current Research:
Conflict Studies; National Security & Nuclear Issues; Ethnicity & Political
Violence.

Main Work at Hand:
Identity, Political Movements & Violence in Jammu & Kashmir; Bridging the
Gap: Confidence Building Measures Between India & Pakistan.

Recent/Forthcoming Publications:
"Enemy Images: The Media & Indo-Pakistan Tension, in Michael Krepon & Amit
Sevak (eds.) *Crisis Prevention Confidence Building & Reconciliation in South
Asia*, (New York, 1995); "Linkages Between Domestic Politics and Diplomacy: A
Case of India & Pakistan", B. Ghosal, ed., (New Delhi, Konark Publishers, 1996);
Identity, Political Movements & Violence in Jammu & Kashmir.

CHARI, Padmanabma Ranganath (M/60)
Co-Director, Institute of Peace & Conflict Studies
11 Vasant Enclave, New Delhi 110057, India.
Tel: 91-11-3388155(W) 6884805(H)
Fax: 3388155(W)
E-mail: ipcş@del2.vsnl.net.in

Specialisation : Chemistry.

Areas of Current Research:
International Security; Non-Proliferation/Disarmament Issues;
India's National Security Problems.

Main Work at Hand:
Indo-US Relations: Nuclear (Missile Technology) Dimensions; Fifty Years of
India's Independence: Evolution Prospects.

Recent/Forthcoming Publications:
Indo-Pak Nuclear Standoff: Role of United States; *Brasstacks & Beyond:
Perceptions & Misperceptions in Indo-Pak Relations*; *Nuclear Non-prolifera-
tion in India and Pakistan: South Asian Perspectives*.

CHELLANEY, Brahma (M/38)
The Centre for Policy Research, Dharma Marg, Chanakyapari
New Delhi 110021, India.

Areas of Current Research:
Export Controls & National Security; Missile Technology Control
Regime; International Security & Weapons of Mass Destruction.

Main Work at Hand:
National Security & Technology Controls; Nuclear & Missile Nonproliferation Strategies of the United States.

Recent/Forthcoming Publications:
Stopping the Bomb: A Study of US Policy.

CHITRAPU, Uday Bhaskar (M/44)
Senior Fellow, Institute for Defence Studies & Analysis
Sapru House, Barakhamba Road, New Delhi 110001, India.
Tel: 91-11-332-5840(W)
Fax: 91-11-332-1851(H)

Specialisation : Electrical Engineering.

Areas of Current Research:
Indian Ocean in the Post Cold War Era; Indo-US Relations; China/Burma.

Main Work at Hand:
Monograph on Indian Ocean; Monograph on Myanmar/China.

Recent/Forthcoming Publications:
"CBM in Indian Ocean, in Maritime Security", *State of Naval & Maritime History*.

CHOPRA, Hardev Singh (M/63)
Professor, West European Studies Division, School of International Studies Jawaharlal Nehru University, New Delhi 110067, India.
Tel: 91-11-667676 Ext.226(W)
Fax: 91-11-6865886(W)

Specialisation : European Studies.

Areas of Current Research:
Processes of Regional Integration in Europe: European Union; Foreign Policies of France & Germany; European Union & SAARC.

Main Work at Hand:
National Identity & Regional Integration/Co-operation: Experience of Europe, Relevance to South Asia; The Yugoslav Break up & European Security; Role of Europe in the Global Order.

Recent/Forthcoming Publications:
The France-German Reconciliation: Relevance to the Prospect of South Asian Peace.

DASGUPTA, Sunil (M/27)
Senior Correspondent, India Today, Living Media India Limited F14/15 Connaught Place, New Delhi 110001, India.
Tel: 91-11-331-5801(W)
Fax: 91-11-331-6180(W)
E-mail: delhi.lmi@axcess.net.in

Specialisation : Commerce.

Areas of Current Research:
South Asian Strategic & International Affairs.

Main Work at Hand:
Stories for India Today.

Recent/Forthcoming Publications:
Economic Engagement in South Asia: Towards a Comprehensive Nonproliferation Policy.

DESAI, Bharat (M/35)
Asst. Professor, International Legal Studies Division,
School of International Studies, Jawaharlal Nehru University,
New Delhi 110067, India.

Tel: 91-11-662460(H)
Fax: 91-11-6865886(W)

Specialisation : Environmental Law.

Areas of Current Research:
International Law; International Environmental Law; Human Rights &
Environment.

Recent/Forthcoming Publications:
"Regional Measures for Environment Protection: A SAARC Initiative",
Yearbook of International Environmental Law, Vol.8, 1991; "Threats to the
World Eco-System: A Role for Social Scientists", *Social Science & Medicine,*
Vol.35, No.4, 1992; "The Bhopal Gas Leakage Litigation", *Asian Yearbook of
International Law,* Vol.3, 1994.

DIPANKAR, Binerjee (M/54)
Co-Director, Institute of Peace & Conflict Studies
11 Vasant Enclave, New Delhi 110057, India.

Specialisation : Political Science, Defence Studies.

Areas of Current Research:
South Asian Security; Proliferation of Small Arms; Nuclear Non-Proliferation.

Main Work at Hand:
Indian Foreign Policy in the New World Order; India - Ethnic Relations.

GUPTA, Ruchira (F/31)
BBC, Aifacs Building,
1 Rafi Marg, New Delhi, India.

Specialisation : English Litrature.

Areas of Current Research:
Religious Violence; Violence Against Women and Children; Facism.

Main Work at Hand:
The Religious Right & Urban Terrorism.

Recent/Forthcoming Publications:
Biography of Mangla Devi Singh.

GUPTA, Dipankar (M/46)
Chairman, Centre for the Study of Social Systems
School of Social Sciences, Jawaharlal Nehru University
New Delhi 110067, India.
Tel: 91-11-667676 Ext 431(W)
Fax: 91-11-686-5886(W)

Specialisation : Sociology

Areas of Current Research:
Ethnicity & Politics; Farmer's Movements; Anthropology of Medicine.

Main Work at Hand:
The Context of Ethnicity: Sikh Identity in a Comparative Perspective.

Recent/Forthcoming Publications:
Rivarly & Brotherhood: Rural Unionism in Uttar Pradesh; Engaging With Events: Political Sociology in India; *Nativism in Metropolitics: The Shiv Sena in Mumbai*, (Manohar, 1982, Delhi).

JAIN, Randhir B. (M/62)
Professor of Political Science, University of Delhi
New Delhi 110007, India.
Tel: 91-11-7257472(H)

Specialisation : Political Science & Public Administration.

Areas of Current Research:
Development Strategies in South Asia; South Asian Co-operation, South Asian Politics & Public Administration.

Main Work at Hand:
Environmental Stewardship & Sustainable Development: South Asian Perspective; Public Service Neutrality: A Comparative Analysis.

Recent/Forthcoming Publications:
NGOs in Development Perspective, (Delhi, Vivek Prakashan, 1995); *Structural Adjustment, Public Policy & Bureaucracy in Developing Societies*, (New Delhi, Har-Anand Publications, 1994); *Bureacracies & Developmental Policies in the Third World*, (Amsterdam, Free University Press, 1993).

KASTURI, Bhashyam (M/31)
The Decan Herald, 6474 Sector B, Pocket 9
Vasant Kunj, New Delhi 110070, India.
Tel: 91-11-6867339(W)
Fax: 91-11-6862077(W)

Specialisation :
South Asian Studies; Strategic & Security Studies; Military History.

Areas of Current Research:
Gandhi & Partition.

Main Work at Hand:
Intelligence & Foreign Policy: The Indian Experience.

Recent/Forthcoming Publications:
Intelligence Services, (Lancer 1995); *Walking Alone: Gandhi & India*, (1997).

KESAVAN, K.V. (M/56)
Professor & Chairman, Centre for East Asian Studies
School of International Studies, J.N.U., New Delhi, India.

Specialisation : International Relations.

Areas of Current Research:
Japanese Foreign Policy; Japanese Domestic Politics; Asia-Pacific Region &
Economic Cooperation.

Main Work at Hand:
Japan's Policies in Indo-China; Japan-US Relations in the Nuclear Field.

Recent/Forthcoming Publications:
Japan's Relations with Southern Asia 1952-60; Japan's Defence Policy Since
1976.

KUMAR, Kishore (M/39)
Society for Indian Ocean Studies, Secular House, 9/1 Institutional Area
Aruna Asaf Ali Marg, New Delhi 110067, India.
Tel: 91-11-651 2701(W)
Fax: 91-11-696 8171(W)
E-mail:sios.delhi@sm.1.sprint.rpg.spm

Specialisation : India's Maritime Strategy.

Areas of Current Research:
South Asia: Politics, Economy & Strategy; Indian Ocean: Strategy & Political Economy; Science Policy.

Main Work at Hand:
India's Oceanic Studies; Journal of Indian Ocean Studies (Assistant Editor)

Recent/Forthcoming Publications:
Dialogue of the Deaf: India-Pakistan Divide, (New Delhi: Konark, 1992);
Defence Versus Development: India Case Study, (New Delhi: Indus, 1993).

KUMAR, Sumita (F/25)
Institute for Defence Studies & Analyses, Sapru House
Barakhamba Road, New Delhi 110001, India.
Tel: 91-11-3321840(W) 6853476(H)
Fax: 91-11-3321851(W)
E-mail: postmast@idsa.delnet.ernet.in

Specialisation : International Relations.

Areas Of Current Research:
South Asian Security; Narco Terrorism In South Asia; Pakistan & Human Rights.

Main Work At Hand:
End of the Cold War: Impact on Pakistan's Security; Cross-Border Terrorism as a Threat to International Security.

Recent/Forthcoming Publications:
"Menace of Drug Trafficking", *World Focus,* (Oct-Dec 1995); "Drug Trafficking in Pakistan", *Asian Strategic Review,* (1994-95); "Pakistan's Security After the Cold War: The US Factor", *Strategic Analysis.*

KUMAR, Sushil (M/58)
Professor of International Politics, School of International Studies
Jawaharlal Nehru University, New Delhi 110067, India.
Tel: 91-11-6106086(H)
Fax: 91-11-686-5886(H)

Specialisation : Political Science.

Areas of Current Research:
Identity, Ethnicity & Political Development; State in Post-Cold War International Relations; South Asian Security & International Relations.

Main Work at Hand:
Security in South Asia; Nationalism & Inter-State Relations in South Asia.

Recent/Forthcoming Publications:
New Globalism & the State: Towards Post-Cold War IR Theory, (New Delhi: Research Press 1996); *Diversity, Order & Justice, Liberal-Realist Dialogue in IR Theory*, (1995); *Gorbachev's Reforms & International Change*, (New Delhi: Lancer, 1993).

LAMA, Mahendra P. (M/34)
Associate Professor, South Asia Division, School of International Studies
Jawaharlal Nehru University, New Delhi 110067, India.
Tel: 91-11-618-8817(H)
Fax: 91-11-618-7435
E-mail: mpl@jnuniv.ernet.in (W)

Specialisation : Development Economics.

Areas of Current Research:
Economic Relations, Bilateral & Multilateral, in South Asia; Trade, Aid, Investment, Technology; Regional Economic Cooperation in South Asia; Environmental Security & Economic Resources in South Asia.

Main Work at Hand:
Economics of Indo-Nepalese Cooperation: A Study on Trade, Aid & Private Foreign Investment; Regional Cooperation in South Asia: Problems, Potentialities.

Recent/Forthcoming Publications:
Indian & Chinese Aid to Nepal; Science & Technology Cooperation in South Asia; Environmental Security & Economic Resources in South Asia.

MANSINGH, Surjit (F/59)
Centre for International Politics, School of International Studies
Jawaharlal Nehru University, New Delhi.

Specialisation : International Studies.

Areas of Current Research:
Comparative Foreign Policy, China, India; Strategic Studies/Conflict Resolution, especially South Asia; International Relations Theory.

Main Work at Hand:
India's Search for Power; Historical Dictionary of India.

Recent/Forthcoming Publications:
Indian & Chinese Foreign Relations in Comparative Perspective, Radiant 1997; The US, India and China World Affairs.

MEHRA, Ajay K. (M/42)
Centre for Policy Research, Dharma Marg
Chankyapuri, New Delhi 110021, India.
Tel: 91-11-301-5273(W)
Fax: 91-11-301-2746(W)

Specialisation : Political Science.

Areas of Current Research:
National Security; Ethnic Conflict; Governance.

Main Work at Hand:
Governance & Accountability: Standing Committees of the Indian Parliament; Police in Changing India.

Recent/Forthcoming Publications:
The Politics of Urban Redevelopment: A Study of Old Delhi, (New Delhi, 1991); *Strengthening Indian Voluntary Action*, (New Delhi, Konark, 1995); *The Indian Cabinet: A Study in Governance*, (Kornak, New Delhi).

MEHTA, Rajesh (M/40)
ICRIER, East Court, Fourth Floor
India Habitat Centre, Lodi Road, New Delhi 110003, India.

Tel: 91-11-4616329(W) 7430350(H)
Fax: 91-11-4620180(W)
E-mail: root@icrier.ren.nic.in

Specialisation : Economics.

Areas of Current Research:
International Economics; Regional Blocs; Indian Economy.

Main Work at Hand:
Regional Trading Blocs; Economic Reforms & Bilateral Trading Arrangements.

Recent/Forthcoming Publications:
Indian Ocean Rim Initiatives.

MUKHERJI, Indra Nath (M/52)
South Asian Studies Division, School of International Studies
Jawaharlal Nehru University, New Delhi 110067, India.

Specialisation : Economics.

Areas of Current Research:
Trade & Development Issues (South Asia); SARRC/SAPTA; WTO.

Main Work at Hand:
Project on SAPTA; India's Bilateral Economic Relations with South Asian
Countries.

Recent/Forthcoming Publications:
Assessing Trade Flows in Negotiated Products in the First Round of Trade
Negotiations, SAARC Chamber of Commerce; Economic & Political Atlas of
SAARC.

MULAY-PARAKH, Regina (F/39)
Centre for International Politics, School of International Studies
Jawaharlal Nehru University, New Delhi 110067, India.
Tel: 91-11-2241801(H)
Fax: 91-11-686-5886(H)

Specialisation : International Politics.

Areas of Current Research:
SAARC Region; Communication; Strategic Studies, Conflict Research.

Main Work at Hand:
International Communication as Tool of Beneficial Co-operation: Study of
SAARC Audio Visual Exchange Programme; Alternatives to Conflict Reso-
lutions.

NARANG, A.S. (M/48)
School of Social Sciences, Indira Gandhi National Open University
New Delhi 110068, India.
Tel: 91-11-6961845(W) 91-11-7450361(H)
Fax: 91-11-6862312(W)

Specialisation : Political Science.

Areas of Current Research:
Ethnic Studies; Regional Conflict; South Asia.

Main Work at Hand:
Politics of Terrorism; Social Basis of Federalism.

Recent/Forthcoming Publications:
Ethnic Identities & Federalism, (Shimla, Indian Institute of Advanced
Studies 1995); *International Terrorism*, Delhi, (Indian Institute of Non-
Aligned Studies, 1993); Social Basis of Federalism.

NORBU, Dawa (M/45)
School of International Studies, Jawaharlal Nehru University
New Delhi 110067, India.
Tel: 91-11-667676 Ext 276(W)
Fax: 91-11-686-5886(W)

Specialisation : Political Science.

Areas of Current Research:
Theories of Non-Western Nationalism; Strategic Developments in Inner Asia;
History, Culture & Sociology of Tibet.

Recent/Forthcoming Publications:
Roots of Nationalism in Post Communist Societies; China's Tibet Policy;
Culture & the Politics of Third World Nationalism; Red Star Over Tibet,
(Collins, 1974, 1987, 1996).

OOMMEN, T.K. (M/58)
Centre for the Study of Social Systems, School of Social Sciences
Jawaharlal Nehru University, New Delhi 110067, India.
Tel: 91-11-6864242(H)
Fax: 91-11-6865886(W) 91-11-6864242(H)

Specialisation : Sociology.

Areas of Current Research:
Social Movements; Political Sociology; Cultural Pluralism; Nation Building.

Recent/Forthcoming Publications:
Alien Concepts & South Asian Reality, (Sage, 1995); Citizenship, Nationality
& Ethnicity: Reconciling Competing Identities; Citizenship & National

Identity: Case Studies from Five Continents; *Social Structure & Politics, Studies in Independant India*, (Hindustan Publishing Corporation, New Delhi, 1994); *State & Society in India: Studies in Nation Building*, (Sage, New Delhi. 1990).*

RAGHAVAN, Sudha (F/51)
Centre for International Politics, Organization and Disarmament
School of International Studies, Jawaharlal Nehru University
New Delhi 110067, India.
Tel: 91-11-667676(W); 573-7328(H)
Fax: (91-11)6962292(W)

Specialisation : South East Asia Area Studies.

Areas of Current Research:
South Asia; South East Asia; South Pacific.

Recent/Forthcoming Publications:
Indian Ocean Power Politics: Attitude of South East Asian & South Pacific Countries, (Lancer Publications, New Delhi, 1996).

RAMAKRISHNAN, Nitya (F/34)
66, Babar Road, New Delhi 110021
India.
Tel: 91-11-3714531(W)

Specialisation : Law.

Areas of Current Research:
Human Rights Related Law in Theory and Practice; Colonial Impact in Third World.

RASGOTRA, M. (M/71)
Rajiv Gandhi Foundation, Jawahar Bhavan
Dr. Rajendra Prasad Road, New Delhi 110001, India.
Tel: 91-11-375 5117(W) (91-11)687 3601(H)
Fax: 91-11-375 5119(W)

Specialisation : International Relations.

Areas of Current Research:
South Asia; Indian Foreign Policy & International Relations;
Nuclear Disarmament.

Main Work at Hand:
Indian Foreign Policy Under Rajiv Gandhi.

Recent/Forthcoming Publications:
Indian Foreign Policy & Diplomacy.

ROY CHAUDHURY, Rahul (M/31)
Research Officer, IDSA, Sapru House,
Barakhamba Road, New Delhi 110001, India.
Tel: 91-11-3321840(W) 225572(H)

Specialisation : International Relations.

Areas of Current Research:
Naval Affairs; Maritime Security in the Indian Ocean; Indian Military Forces.

Recent/Forthcoming Publications:
"India's Security Policy", *Strategic Analysis,* May 1996; "Defence Alert",
Seminar, Jan 1996; "Trends in Major Naval Arms Transfers in Indian
Ocean", *Strategic Analysis,* March 1996; *Sea Power & Indian Security*,
(Brassy's, London, 1995); "Defence Research & Development in India",
Strategic Review 1994-1995.

SACHDEVA, Gulshan (M/30)
Centre for Policy Research, Dharma Marg
Chanakyapuri, New Delhi 110021, India.
Tel: 91-11-3015273(W)
Fax: 91-11-3012746(W)
E-mail: manager@cpr.delnet.ernet.in

Specialisation : Economics.

Areas of Current Research:
Liberalisation in the Developing World; Trade Related Issues.

Main Work at Hand:
Economic Development of the North-Eastern Part of India; Indian Ocean
Economic Cooperation.

Recent/Forthcoming Publications:
"Liberalisation & the North-East (of India)", *Mainstream,* Vol.33, No.47,
1995.

SAHADEVAN, P. (M/34)
South Asian Studies Division, School of International Studies
Jawaharlal Nehru University, New Delhi 110067, India.
Tel: 6167676 EXT 276(W)
E-mail:saha@jnuniv.ernet.in

Specialisation : South Asian Studies.

Areas of Current Research:
Domestic Politics & Foreign Policy of Sri Lanka; Ethnic Conflict & it's
Resolution.

Main Work at Hand:
India's Foreign Policy & Relations, 1972-95.

Recent/Forthcoming Publications:
India & Overseas Indians: The Case of Sri Lanka, (New Delhi, 1995); *Lost Opportunities & Changing Demands: Explaining the Ethnic Conflict in Sri Lanka*, (Kent Paper, *Canterbury*, 1995); "On Not Becoming a Democrat: The LTTE's Commitment to Armed Struggle", *International Studies*, (New Delhi, Oct-Dec 1995.

SINGH, K.R. (M/63)
School of International Studies, J.N.U.,
New Delhi 110067, India.
Tel: 91-11-667676 Ext 384(W) 91-11-6965656(H)

Specialisation : International Studies.

Areas of Current Research:
Middle East & North Africa; Indian Ocean; National Security and Strategic Studies.

Main Work at Hand:
Evolution of Maritime Strategy in South Asia.

Recent/Forthcoming Publications:
Intra Regional Interventions in South Asia, (United Service Institution of India, New Delhi 1990); *Making of an Indian Ocean Community*, (New Delhi 1995).

SINGH, Swaran Jaswal (M/34)
I.D.S.A., Sapru House
Barakhamba Road, New Delhi 110001, India.
Tel: 91-11-689-9052(H)
Fax: 91-11-332-1851(W)

Specialisation : Limited War.

Areas of Current Research:
China's Defence & Modernisation; Indian Security/Defence; East Asia.

Main Work at Hand:
China's Changing National Security Strategy; India's Higher Defence Organization.

Recent/Forthcoming Publications:
Limited War: The Challenge of US Military Srategy, (New Delhi: Lancer's Books, 1995).

SINHA, P.B. (M/56)
Institute for Defence Studies & Analyses, Sapru House
Barakhamba Road, New Delhi 110001, India.

Specialisation : History.

Areas of Current Research:
International Terrorism; South Asia; Nuclear Politics.

Main Work at Hand:
Nuclear Pakistan: Atomic Threat to South Asia; Armed Forces of Bangladesh.

Recent/Forthcoming Publications:
"Islamic Fundamentalism & Terrorism in Algeria", *Strategic Analysis,* June 1994; "Continuing Violence in Karachi", *Strategic Analysis,* April 1995; "Muslim Insurgency in the Philippines", *Strategic Analysis,* Aug 1995.

SINHA, Prabhas Chandra (M/35)
POG /CIPOD, 319 SIS
Jawaharlal Nehru University, New Delhi 110067, India.
Tel: 91-11-667676(W) (91-11)6899877(H)
Fax: 91-11-6899534(H)

Specialisation : Political Geography.

Areas of Current Research:
Oceans & Antarctic Strategy; Environmental Security; Political Geography.

Main Work at Hand:
Third World Perspective on Oceans & Antarctic; International Environmental Security.

Recent/Forthcoming Publications:
India's Ocean Policy; Third World Perspective on Arctic & Antarctic; Conflict Resolution: A Gandhian Approach.

SIVARAMAKRISHNAN, Siddhartha (M/23)
Ford Foundation
55, Lodi State, New Delhi 110003
India.
Tel: 91-11-4619441(W)
Fax: 91-11-4627147(W)
E-mail: s.krishnan@fordfound.org

Specialisation : Public Administration.

Areas of Current Research:
Normative Dimensions of Public Policies; South Asian Regional Cooperation.

Main Work at Hand:
Regional Security & Cooperation; Local Governance.

Recent/Forthcoming Publications:
"The Individual & the Group in International Human Rights", *Maxwell Colloquium,* Spring 1996; "National Planning & Political Liberties", *Maxwell Review,* Spring 1995.

SRIDHARAN, Eswaran (M/39)
Centre for Policy Research, Dharma Marg
Chanakyapuri, New Delhi 110021, India.
Tel: 91-11-3015273(W) 91-11-6873063(H)
Fax: 91-11-3012746(W)

Specialisation : Political science.

Areas of Current Research:
Political Economy of Development; Political Economy of International Relations; Emerging Asia Pacific, Japan, China, ASEAN.

Main Work at Hand:
Party System Change, Coalition Theory & Economic Performance in India; Political Economy of the Emerging Asia-Pacific Basin.

Recent/Forthcoming Publications:
The Political Economy of Industrial Promotion: Indian, Brazilian & Korean Elections in Comparative Perspective 1969-1994.

SREEDHARA, T. Rao (M/54)
Institute of Defence Studies & Analyses, Sapru House,
Barakambha Road, New Delhi 110001, India.

Specialisation : Economics.

Areas of Current Research:
South Asia; Persian Gulf.

Main Work at Hand:
Trends in India's Defence Expenditure Since Independence.

Recent/Forthcoming Publications:
"Iranian Armed Forces", A Status Report, Feb '96; "Strategic Environment of Iran", June '96.

THAPLIYAL, Sangeeta (F/29)
Institute for Defence Studies & Analyses, Sapru House
Barakhamba Road, New Delhi 110001, India.

Specialisation : Political Science.

Areas of Current Research:
Strategic Concerns of South Asian Countries; Security Concerns of Small States; Confidence Building Measures as a Tool for Regional Security & Stability.

Main Work at Hand:
India's Security Arrangements with Nepal; Role of Superpowers in Regional Conflicts - A Study of Indo-Pakistan Relations.

Recent/Forthcoming Publications:
Mahakali - Bridging the Gap; Gurkhas in the Indian Army; Water & Conflict.

VISVANATHAN, Shiv (M/45)
Centre for Study of Developing Societies, 29, Rajpur Road
New Delhi 110054, India.

Specialisation : Sociology.

Areas of Current Research:
Sociology of Science; Literature as Political Theory; Democracy as a Heuristic.

Recent/Forthcoming Publications:
"Re-Thinking Rights", "Three Tables in Communication Theory", *IIC Quarterly* 1996; The Fatal Concept: Re-Thinking Security; *Organizing for Science* (OUP, 1985); *A Carnival for Science* (OUP, 1996).

WARIAVWALLA, Bharat K. (M/62)
Centre for the Study of Developing Societies
29 Rajpur Road, New Delhi 110054, India.

Areas of Current Research:
State, Nation & Nation-State Problematique; Changing Role of the State in the Global Political Economy; Human Rights.

Main Work at Hand:
State & the Multi-Ethnic Socities of the Developing World.

Recent/Forthcoming Publications:
Demolition & the Nation-State Problamatique, (University of Lepzig Publication); "Security with Status", *AAAS Annual*, 1993; Nationalism & Nation Making.

WARIKOO, K. (M/44)
Associate Professor, Central Asian & Himalayan Studies
School of International Studies,
Jawaharlal Nehru University, New Delhi 110067, India.
Tel: 91-11-6862763(H)
Fax: 91-11-6859408(H)

Specialisation : Central Asian Studies.

Areas of Current Research:
Central Asian History (Modern), Society, Culture & Politics; Geopolitics of Central Asia; Kashmir Problem - Domestic & External Dimensions.

Main Work at Hand:
Ethnic - Religious Diversity in Jammu & Kashmir.

Recent/Forthcoming Publications:
Society & Culture in the Himalaya (ed.), (New Delhi, Har Anand, 1995); *Central Asia: Emerging New Order* (ed.), (New Delhi, Har Anand, 1995); *Jammu, Kashmir & Ladakh: Linguistic Predicament* (ed.), (New Delhi, Har Anand, 1996).

■ Patna (Bihar)

LAL, Ravindra Kumar (M/55)
Dept. of Political Science, Patna University
Patna 800005, India.

Specialisation : Political Science.

Areas of Current Research:
International Relations; Public Administration.

Main Work at Hand:
Indian Foreign Policy & Relations.

■ Pondicherry (TN)

JHA, N.K. (M/36)
Associate Professor, School of International Studies
Pondicherry Central University, Pondicherry 605014, India.
Tel: 91-413-85-2265(W)

Specialisation : India's Foreign Policy.

Areas of Current Research:
Domestic Imperatives in India's Foreign Policy & Indo-American Relations; International Relations Theory; Foreign Policy of Bangladesh.

Main Work at Hand:
India-Bangladesh Relations: Political Dimensions; Domestic Compulsions in India's US Policy.

Recent/Forthcoming Publications:
Non-Alignment & Nation Building: The Indian Experience; India's Nuclear Policy.

■ Pune (Maharashtra)

PARANJPE, Shrikant (M/41)
Dept. of Defence & Strategic Studies, University of Pune
Pune 411007, India.

Specialisation : Political Science.

Areas of Current Research:
South Asian Security Issues; Nuclear Policies.

Recent/Forthcoming Publications:
Parliament & Foreign Policy: A Study of Indian Nuclear Policy, ICSSR
Project Report, (Radiant Publishers, New Delhi); *India & South Asia Since
1971*, (New Delhi:Radiant, 1985); *US Non Proliferation Policy in Action:
South Asia*, (New Delhi: Sterling, 1987).

■ Ranchi (Bihar)

LAL, Mani Prasad (M/48)
Dept. of Political Science, Marwani College
Ranchi, India.
Tel:834012(H)

Specialisation : Political Science.

Areas of Current Research:
Tribal Administration in India; Tribal Politics in India; Panchayati Raj.

Main Work at Hand:
Problems of Land in the District of Ranchi.

Recent/Forthcoming Publications:
"Ethics in Tribal Administration", *Indian Journal of Public Administration*,
July-Oct 1995; Panchayati Raj in Bihar.

SINDHU, V. (F/24)
12-8-317/A, Alugadda - Bavi
Sec'bad, Andhra Pradesh, India.

Specialisation : International Relations.

Areas of Current Research:
South Asia; U.S. Nuclear Policy; Peace Studies.

Main Work at Hand:
India's Strategic & Security Perspective.

■ *Srinagar (J & K)*

MATTOO, Abdul Majid (M/50)
Director, Central Asian Studies
Kashmir University, Srinagar 19006, India.
Fax:0194-78202(W)

Specialisation : History.

Areas of Current Research:
Contemporary Central Asia; Medieval Turkistan, Indian History & Culture;
Kashmir History.

Main Work at Hand:
Transfer of Men & Movements Along Silk Route; Integrated History of
Turkistan Since the Advent of Islam.

Recent/Forthcoming Publications:
Kashmir During the Afghan Rule (1752-1819); Descriptive Catalogue of
Persian Manuscripts Research Library Collection.

■ *Thiruvananthapuram (Kerala)*

MOHANAN, B. (M/39)
ISDA, 4/64-2, Continental Gardens, Kowdiar
Thiruvananthapuram 695003, Kerala, India.
Tel:0471-436101(W)

Specialisation : Regional Cooperation in South Asia

Areas of Current Research:
Regional Cooperation/Integration Studies; Foriegn Policies in South Asia;
Development Studies.

Main Work at Hand:
Globalization & the State; Democracy in South Asia 2000 AD & Beyond.

Recent/Forthcoming Publications:
Post Cold War Perspectives on International Relations; Mahatma Gandhi's
Legacy & the World Order; Indo-U.S. Economic Relations.

■ *Udaipur (Rajasthan)*

AGWANI, Mohammed Ahafi (M/67)
37, Ahimsapuri,, Udaipur 313001, India.

Specialisation : International Affairs.

Areas of Current Research:
International Politics; Modern West Asia; Contemporary Islamic Movements.

Main Work at Hand:
India & the Gulf.

Recent/Forthcoming Publications:
Contemporary West Asia.

MEHTA, Jagat Singh (M/73)
Vidya Bhawan Society, Dr. M.S. Mehta Marg
Udaipur 313001, India.
Tel: 91-94-560311(W) 91-294-524389(H)
Fax:91-294-560047(W)

Specialisation : Economics.

Areas of Current Research:
Roots of Misperception in Cold-War Politics; South Asian Security &
Development; Water Diplomacy in South Asia.

Main Work at Hand:
Rescuing the Future from Misperspection.

■ *Varanasi (UP)*

KHAN, Diwan Ghufran Ahmad (M/44)
Dept. of Political Science, Banaras Hindu University
Varanasi 221005, India.
Tel:545304(W) 91-542-314295(H)

Specialisation : International Relations.

Areas of Current Research:
SAARC; Conflict Resolution; Government & Politics of South Asian Countries.

Main Work at Hand:
1973 Constitution of Pakistan: A Study of the Office of the Prime Minister;
India as a Factor in British Policy Towards South Asia in Post Cold War
Period.

Recent/Forthcoming Publications:
Government & Politics of Pakistan, (Naya Sansar Press, Varanasi, 1996).

■ *Visakapatnam (AP)*

TRIPURANA, Nirmala Devi (F/44)
Reader, Centre for SAARC Studies, Kirlampudi Area, Andhra University
Visakhapatnam 530017, Andhra Pradesh, India.
Tel:575328(W)

Specialisation : International Economics.

Areas of Current Research:
South Asian Studies; International Economics; Women's Studies.

Main Work at Hand:
Research Paper on SAPTA.

Recent/Forthcoming Publications:
Population & Development in SAARC, (South Asian Publishers Pvt. Ltd., New Delhi, 1996.)

IRAN

KILORRAM, Ali (M/43)
Institute for Political & International Studies
P.O. Box 19375-1793, Tehran, Iran.
Tel: 2571015/673781(W)
Fax: 270964/674763(W)

Specialisation : Physics.

Areas of Current Research:
Indian Ocean Community; Asian Identity; Persian Gulf's Security.

Main Work at Hand:
Afghanistan Crisis; China Affairs; Indo-Pakistan Tension.

Recent/Forthcoming Publications:
Muslims in China; Asian Identity, an Appropriate Basis for Cooperation in Asia.

TAHERI AMIN, Zahra (F/40)
Institute for Political & International Studies
P.O. Box 19395-1793, Tehran, Iran.
Tel: 2571015/673781(W)
Fax: 270964/674763(W)

Specialisation : Political Science.

Areas of Current Research:
Chinese Affairs, Indian Ocean Community; Women's Affairs in Asia.

Main Work at Hand:
Muslims in China; Indian Ocean Community.

Recent/Forthcoming Publications:
The Chronology of Events in China Since 1949, (Pub. by IPIS in 1990); *The Influence of China in Africa*, (Pub. by IPIS in 1993); *Women's Status in Muslim Countries*, (Pub. by IPIS, 1995).

ISRAEL

KUMARASWAMY, P.R. (M/34)
Research Fellow, Truman Institute
Herbrew University, Jerusalem 91905, Israel.
Fax: 972-2-567 0635(W)

Specialisation : International Studies.

Areas of Current Research:
Middle East Politics; Israel & South Asia; Israel & China.

Main Work at Hand:
South Asia After the Cold War; India & Israel.

LIMAYE, Satu P. (M/32)
Japan Institute of International Affairs, Toranomon Mitsui Building
3F 3-B-1 Kasumigasoki, Chiyoda-Ku, Tokyo 100, Japan.
Tel: 81-3-3504-0302(W)
Fax: 81-3-3595-1755(W)
E-mail: 71124.236@compurserve.com

Specialisation : International Relations.

Areas of Current Research:
US Relations with South Asia; Japan's Relations with South Asia; International Relations.

Main Work at Hand:
US-Indian Relations: The Pursuit of Accommodation, (Westview Press, 1993); "Sushi & Samosas: Indo-Japanese Relations After the Cold War", in Sandy Gordon, *India Looks East*, (Australian National University, 1995).

MALDIVES

FAIZAL, Farahanaz (F)
Ma. Andromeda
Chandani Magu
Male, Maldives.
Tel: 960-316940(W), 326278(H);
Fax: 960-324504

Specialisation: International Relations

Areas of Current Research:
Security of Small States

Recent/Forthcoming Publications:
"Insecurity and Islandness: The Case of Maldives".

JALEEL, Mohamed (M)
Manager, Economic Research and Statistics Division
Maldives Monetary Authority
Male, MALDIVES.
Fax: 960-323862

Area of Specialisation: Economics

Areas of Current Research:
Macro-economics, Development, International Institutions

Recent/Forthcoming Publications:
"Private Sector and Regional Cooperation in South Asia"; "Regional Economic Trends and Security Implications for Maldives" in Iftekharuzzaman ed. *Regional Economic Trends and South Asian Security* (Manohar, Delhi, 1997).

SHAHEED, Ahmed (M)
Director, Foreign Relations
Ministry of Foreign Affairs
Male, Maldives.
Tel: 960-322 640 (W); 322 057 (H)
Fax: 960-323 841.

Specialization: Foreign Policy and Security of South Asia

Areas of Current Research:
Problems of Small States, Developmental Issues and Regional Cooperation.

Recent/Forthcoming Publications:
"Regional Economic Trends and Security Implications for Maldives" in Iftekharuzzaman ed. *Regional Economic Trends and South Asian Security* (Manohar, Delhi, 1997).

ZAKI, Ibrahim H. (M)
Former Secretary General of SAARC
Minister for Tourism
Male, Maldives.
Tel: 960-325027(W)
Fax: 960-322512
e-mail: tourism@dhivehinet.net.mv

Specialisation: Regional Cooperation, Tourism and Development

Areas of Current Research:
Regional Cooperation and Confidence Building in South Asia

Recent/Forthcoming Publications:
"Recent Developments in SAARC and Prospects for the Future" in *South Asian Survey* (Delhi, March-August 1994).

■ *Kathmandu*

ADITYA, Anand (M/51)
NEFAS, Kathmandu, Nepal.
Tel: 977-1-227751(W)

Specialisation : Political Science.

Areas of Current Research:
Research Methodology; Elections; Development Studies.

Main Work at Hand:
Democracy & Empowerment in South Asia, Co-editing with Devodra Ray;
The Political Economy of Small States, (ed.)

BARAL, Lok Raj (M/56)
Dept. of Political Science, Tribhuvan University
Kirtipur, Kathmandu, Nepal.
Tel: 977-1-521522(H), 527629(W)

Areas of Current Research:
South Asian Affairs; Security & Migration; Domestic Policy.

Main Work at Hand:
Regional Migration; Ethnicity & Security; Refugees & Regional Security in
South Asia; Problems of Governance.

CHHETRI, Rakesh Kumar (M/46)
Centre for Protection of Minorities & Against Racism & Discrimination
(CEMARD), P.O. Box 3485, Kathmandu, Nepal.

Specialisation : Political Science.

Areas of Current Research:
Ethnicity & Security; Democracy; Problems of Governance.

Main Work at Hand:
"State, Ethnicity & Democracy: Problems of Governance in Bhutan".

DAHAL, Dev Raj (M/38)
G.P.O. Box 6154, Kathmandu
Nepal.
Tel: 977-1-528464(W)
Fax: 977-1-227751(W)

Specialisation : Political Science.

Areas of Current Research:
Nepalese Foreign Policy; Small States; Development Issues.

Main Work at Hand:
Development Studies: Self-Help Organizations, NGO's & Civil Society;
Peace-Keeping Role of Nepal in the UN.

Recent/Forthcoming Publications:
Nepal & the UN; Development Strategy for Nepal.

DIXIT, Ajaya (M/41)
P.O. Box 2221, Kathmandu
Nepal.
Tel: 977-1-528111(W) 977-1-415748(H)
Fax: 977-1-529080(W)

Specialisation : Hydrology.

Areas of Current Research:
Water Resources Management; Water Environment Interface; Water Education.

Main Work at Hand:
1993 Floods in Nepal.

Recent/Forthcoming Publications:
Natural Disasters & Social Resilience; Water, Power, People (Co-author).

DIXIT, Kanak Mani (M/40)
Editor, Himal
G.P.O. Box 7251, Kathmandu,
Nepal.
Tel: 977-1-52345(W)
Fax: 977-1-521013
E-mail: himalmag@mos.com.np

Specialisation : International Affairs.

Areas of Current Research:
Society; Geopolitics.

Main Work at Hand:
Editor, *Himal, South Asia.*

GYAWALI, Dipak (M/45)
Inter Disciplinary Analysts, G.P.O. Box 3971
Kathmandu, Nepal.
Tel: 977-1-528111(W) 977-1-522786(H)
Fax: 977-1-529080(W)
E-mail: post@ida.wlink.com.np

Specialisation : Hydropower, Political Economy.

Areas of Current Research:
Interdisciplinary - Water, Energy, Society Interface.

Main Work at Hand:
1993 Floods in Nepal, Natural Disaster & Social Reliance; Nepal-India Water Relations.

Recent/Forthcoming Publications:
"Arun III Autopsy", in *Globalism & the South*, (Macmillan 1997).

JAISWAL, Tribhuwan Nath (M/52)
Head of Dept., Central Dept. of Political Science
Tribhuvan University, Kirtipur, Kathmandu, Nepal.

Areas of Current Research:
Foreign Relations Among South Asian Countries; Operations of Political Systems; Disarmament & Arms Control: Recent Trends & Prospects.

Main Work at Hand:
The UN Peace Keeping Activities: Nepal's Role; A Critical Review of Nepal's Relationship with India from 1990-94.

Recent/Forthcoming Publications:
Nepal's Foreign Policy; Nepal-Egypt Relations.

JOSSE, Mana Ranjan (M/54)
People's Review, Pipal Ko Both Dillibazar
Kathmandu, Nepal.
Tel: 977-1-471098(H)

Specialisation : International Relations.

Areas of Current Research:
Foreign and Security Policy of Nepal and South Asia; Nepalese Politics; Media.

Main Work at Hand:
Nepal & the World: An Editor's Note Book - In Two Volumes.

Recent/Forthcoming Publications:
Role of Nepal in UN Security Council, 1988-89; Role of Nonaligned Caucus in UN Security Council, 1988-89.

KHANAL, Krishna P. (M/42)
Centre for Nepal & Asian Studies, Tribhuvan University
Kirtipur, Kathmandu, Nepal.
Tel: 977-1-227184(W)

Specialisation : Political Science.

Areas of Current Research:
Politcal Process; Ethnic Studies; South Asia; South Asian Cooperation.

Main Work at Hand:
People, Polity & Governance in Nepal; Constitutional Development in Nepal.

Recent/Forthcoming Publications:
"Party Politics & Governance: The Role of Leadership; State, Leadership and Politics in Nepal"; Socio-Economic Imperatives of SAARC; "Anti-India Feelings in South Asia", *Nepalese Journal of Political Science,* 1988.

KHATRI, Sridhar K. (M/42)
Central Dept. of Political Science, Tribhuvan University
P.O. Box 10619, Kathmandu, Nepal.
Tel: 977-1-412477(W)
E-mail: sridhar@khatri.wlink.com.np

Specialisation : International Affairs.

Areas of Current Research:
Regional Security in South Asia; Regional Cooperation.

Main Work at Hand:
"A Decade of South Asian Regionalism", *Contemporary South Asia,* Vol.1, No.1, 1990; "Security Issues in South Asia: A Nepalese Perspective", *South Asian Survey,* Vol.2, No.1, 1995.

KUMAR, Dhruba (M/46)
Centre for Nepal & Asian Studies, Tribhuvan University
Kirtipur, P.O. Box 3757, Kathmandu, Nepal.
Tel: 977-1-231740(W) 977-1-421042(H)
Fax: 977-1-227184(W)

Areas of Current Research:
Strategic Studies; Chinese Foreign Policy; South Asian Conflict & Cooperation.

Main Work at Hand:
Nuclear Policies & Strategic Culture in South Asia; China & South Asian Security Strategy.

Recent/Forthcoming Publications:
"Nuclear & Missile Proliferation & Future of Regional Security in South Asia", *BIISS Journal,* Jan 1994; Nuclear Non Proliferation in South Asia - A Nepali Perspective; Proliferation, Deterrence & Security Delusion in South Asia.

LOHANI, Mohan Prasad (M/55)
G.P.O. Box 4006, Kathmandu
Nepal.
Tel: 412336(H)

Specialisation : English Literature.

Areas of Current Research:
Regional Cooperation in South Asia; Security Dimension; Strengthening
SAARC through People-to-People Interaction.

MALLA, Shanker Krishna (M/58)
Chairman, Electricity Tariff Commission
Exhibition Road, P.O. Box 2507, Kathmandu, Nepal.
Tel: 977-1-231407(W) 524446(H)
Fax: 977-1-227537(W)

Specialisation : Electrical Engineering.

Areas of Current Research:
Hydro-Power Engineering; Water Resources Development.

Main Work at Hand:
Regional Cooperation in Water Resources Development; Regional Energy &
Gas Grid.

Recent/Forthcoming Publications:
Water Resources Development - Nepalese Perspective; Converting Water
into Wealth.

MATHEMA, Kedar Bhakta (M/50)
P.O. Box 8211, Kathmandu, Nepal.

Specialisation : Literature.

Areas of Current Research:
Issues in Education System; Inequity in Nepalis Education System; Free
Market & Nepal's Schools.

PANDEY, Posh Raj (M/34)
Central Dept. of Economics, Tribhuvan University
Kathmandu, Nepal.
Tel: 977-1-225550(W) 477-1-426058(H)

Specialisation : Economics.

Areas of Current Research:
Regional Cooperation in Trade, Finance & Industry; Equity.

Main Work at Hand:
Development Strategies in Nepal.

SHARMA, Sudhindra (M/31)
G.P.O. Box 2221, Kathmandu
Nepal.
Tel: 977-1-529080(W) 977-1-523181(H)
Fax: 977-1-528111(W)

Specialisation : Sociology.

Areas of Current Research:
Sociology & Anthropology of Religion; Sociology of Water Resource
Management; Ethnic and Religious Conflict.

Recent/Forthcoming Publications:
"Water, Religion & Development - Strains of Discord", *Water Nepal*, Vol.5,
1992; "How the Crescent Fares in Nepal", *Himal*, Nov-Dec 1994; "How
Hindu is the Other Hindustan", *Himal, South Asia*, May 1996; Socio Cultural, Religious & Educational Factors - Nepali Perspective.

SHRESTHA, Govind Das (M/54)
P.O. Box 5627, Kathmandu,
Nepal.

Specialisation : Economics; International Law.

Areas of Current Research:
Water Resources Development, Legal & Institutional; Development Policy.

KLOOS, Peter (M/59)
Dept. of Anthropology, De Boelelaan 1105
10081, HV Amsterdam, Netherlands.
Fax: 31-02-4446722(W)

Specialisation : Anthropology.

Areas of Current Research:
Causes & Consequences of Political Violence.

Recent/Forthcoming Publications:
"Aspects of Violence in Sri Lanka", *Sri Lanka Studies*, (Amsterdam, No. 5, 1996); "Nationalism & Social Research in Sri Lanka", *Social Anthropology*, 3(2):115-128, 1995.

NEW ZEALAND

DICKENS, David (M/33)
Centre for Strategic Studies, Victoria University of Wellington
P.O. Box 600, Wellington, New Zealand.
Tel: 64-4-4955233(W)
Fax: 64-4-4965434(W)
E-mail: dave.dickens@vuw.ac.nz

Specialisation : Public Adminstration.

Areas of Current Research:
India's Relations with South East & East Asia; India's Long Term Strategic Importance; India's Relations with Pakistan, Sri Lanka, Myanmar & China.

Main Work at Hand:
India's Strategic Significance.

■ *Islamabad*

ABBAS, Zaffar (M/40)
c/o BBC Bureau, H-6, Street 8
F-7/3, Islamabad, Pakistan.
Tel: 92-51-823438(W) 92-51-253129(H)
Fax: 92-51270420(W)
E-mail: zaffar@bbc-isb.sdnpk.undo.org

Specialisation : Physics ; Maths.

Areas of Current Research:
Ethnic Conflicts in South Asia; Sectarian, Communal Conflicts & their
Origin; Causes of Poverty in South Asia.

Main Work at Hand:
Mohajir Phenomenon in Karachi; Articles/Features to *Herald Magazine*,
Karachi.

AFZAL, Mujeeb (M/28)
Area Study Centre Quaid-i-Azam University
Islamabad Pakistan
Tel: 92-51-825809/812546(W)
Fax: 92-51-821393(W)

Specialisation : International Relations.

Areas of Current Research:
Pakistan Foreign Policy; Indian Policy Towards Kashmir; Pakistan's Nuclear
Policy.

Main Work at Hand:
"Karachi Crisis", *Current Affairs*, April, 1996; "US Policy Towards South
Asia", *Journal of American Studies*, Area Study Centre, QAU.

AHMAD, Kamaran (M/33)
HRCP, 1-B, Street 38, F-8/1
Islamabad, Pakistan.
Tel: 92-51-811993(H)
E-mail: kamran@dervish.sdnpk.undp.org

Specialisation : Philosophy & Religion.

Areas of Current Research:
Religion, Tolerance/Spirituality; Men's Issues & the Feminist Movement;
Social Activism.

AHMAD, Khurshid (M/63)
Institute of Policy Studies, Block 19, Markat F-7
Islamabad, Pakistan.
Tel: 92-51-818230(W) 051-270869(H)
Fax: 92-51-824704(W)

Specialisation : Economics.

Areas of Current Research:
Islamic Economics; Muslim World; International Relations.

Recent/Forthcoming Publications:
"Islamic Resurgence: Challenges, Directions & Future Perspective - A
Roundtable with Prof. Khurshid Ahmad", (Institute of Policy Studies,
Islamabad, Pakistan, 1995); "Islamic Approach to Development", (Institute
of Policy Studies, 1994); "Nizam-e-Ta'aleem", (Institute of Policy Studies,
1993).

AHMED, Samina (F/43)
Consultant
The Asia Foundation,
P.O. Box 1165, Islamabad 44000, Pakistan.
Tel: 92-51-270590(W)
Fax: 92-51-275436(W)

Specialisation : Political Science.

Areas of Current Research:
South Asian Defence & Strategic Analysis; Civil Military Relations in South
Asia; Pakistani Politics & Foreign Policy.

Main Work at Hand:
The Crisis of State Building & Legtimacy in Afghanistan, Regional Studies;
Ethnicity & Governance in Pakistan.

Recent/Forthcoming Publications:
"Pakistani Conceptions of Security", Muthiah Alagappa ed., *Asian Concep-
tions of Security*, (Stanford University Press); with David Cortright (ed.),
Pakistani Nuclear Options, (Notre Dame Press / Stanford University Press);
"The Political Implications of Ethnicity in Central Asia", *Regional Studies*,
Vol.13, No.2, Spring 1995.

AKHTAR, Safir (M/48)
Institute of Policy Studies, Nasr Chambers, Block 19
Markaz F-7, Islamabad, Pakistan.
Tel: 92-51-818230(W)
Fax: 92-51-824704(W)

Specialisation : Political Science.

Areas of Current Research:
Religions and Politics in South Asia; Political Development in Central Asia.

Main Work at Hand:
Ahl-i-Hadith of Pakistan; Editing "Alam-i-Islam Awr Isaiyyat" Monthly.

AKHTAR, Shaheen (F/35)
Research Analyst, Institute of Regional Studies, NAFDEC Complex 56-F,
Blue Area, Nazimudin Road, F-6/1, Islamabad, Pakistan.
Tel: 92-51-9204934(W) 540516(H)
Fax: 92-51-9204055(W)

Specialisation : International Relations.

Areas of Current Research:
Security & Strategic Issues in South Asia; Ethnicity & Autonomy Movements
in the South Asian Region; Foreign Policy Behaviour of the South Asian States.

Main Work at Hand:
Peace Process in Sri Lanka: Problems & Prospects; Indian Lok Sabha
Elections 1996.

Recent/Forthcoming Publications:
"Determinants of Foreign Policy Behaviour of Small States in South Asia",
Regional Studies, Spring, 1995; "The State of Muslims in India", *Spotlight,*
Regional Affairs, Feb-Mar 1996; Geo-Strategic Dynamics of Indo-Sri Lanka
Relations.

ALI, Mahdi (M/24)
Research Fellow, Human Development Centre
8, Baznar Road, G-6/4, Islamabad, Pakistan.

Specialisation : Economics.

Areas of Current Research:
Defence Studies; Poverty; Debt.

Recent/Forthcoming Publications:
Contribution to Human Development Report, South Asia.

ALI, Salamat (M/62)
The Muslim, Abpara Market
Islamabad, Pakistan.
Tel: 92-51-277484(W) 274835(H)
Fax: 92-51-277485(W)

Specialisation : Chemistry; Zoology.

Areas of Current Research:
International Relations; Economics; South Asian Affairs.

ANIS, Ahmad (M/52)
Dawah Academy, International Islamic University
P.O. Box No.1485, Islamabad, Pakistan.

Specialisation : Comparative Religion.

Areas of Current Research:
Gender Studies; Inter-Faith Dialogue; Islam & Contemporary Issues.

Main Work at Hand:
Women & Social Justice; Islam & the World Religions.

Recent/Forthcoming Publications:
"Ramadan", *Oxford Encyclopedia of Islam & the Muslims*, (New York 1994);
"Abdul A'la Maududi", *Encyclopedia of Islam*, (Islamabad 1994); Women &
Social Justice in Islamic Perspective.

ASHRAF, Fahmida (F/32)
Institute of Strategic Studies, Sector F-5/2
P.O. Box 1173, Islamabad, Pakistan.
Tel: 92-51-9204423(W)
Fax: 92-51-9204658(W)

Specialisation : International Relations.

Areas of Current Research:
South Asia - Political Developments / Policies.

Recent/Forthcoming Publications:
"Kashmir Dispute & India's Intransigence", *Strategic Studies,* Vol.13, Winter
95 & Spring 96, No.384.

AZAM, R.M. Ikram (M/54)
Chairman, Pakistan Futuristic Foundation & Institute,
House No.37, Street No.33, F-8/1, Islamabad, Pakistan.
Tel: 92-51-262116(W) 9204359(H)

Specialisation : International Relations.

Areas of Current Research:
Futuristics: Pakistan, Islam, The Muslim/Third World & Global Human
Futures; The 21st Century New Education; Social Change, Perennial Values
& Life Styles.

Main Work at Hand:
Srategic Futurization: Planning for the 21st Century Futures; Islami Futurism
& Futuristics.

Recent/Forthcoming Publications:
Pakistan & the 21st Century New Human Order, (PFI, 1994); *The 21st Century New Education & the Basic Life Skills*, (PFI, 1996); Muhammad: The Sublime Futurist; Islamic Meditation & Psychotheraphy.

BANURI, Tariq (M/46)
46, Street 12, F-6/3, Islamabad, Pakistan.
Tel: 92-51-278134(W)
Fax: 92-51-278135(W)

Specialisation : Economics.

Areas of Current Research:
Environmental Economics; Governance; Culture & Development.

CHAUDHRY, M. Ghaffar (M/53)
Pakistan Institute of Development Economics,
P.O. Box 1091, Islamabad 45320, Pakistan.
Tel: 92-51-211859(W) 280558(H)
Fax: 92-51-210886(W)
E-mail: arshad%pide@sdnpk.undp.org

Specialisation : Agricultural Econnomics.

Areas of Current Research:
Agriculture & Rural Development; Public Finance & Policy; Institutional & Labour Economics.

Main Work at Hand:
Land Perform in Pakistan: A Historical Perspective; Household Food Security in Pakistan with Reference to the Ration Shop System.

Recent/Forthcoming Publications:
"Transformation of Agriculture, Food Self-Sufficiency & Prospects for Surpluses: The Case of South Asia", *Journal of Contemporary South Asia*, Vol.3, No.1.

CHEEMA, Pervaiz Iqbal (M/55)
Iqbal Fellow, South Asia Institute, Universitat, Heidelberg
Im Neuenheimer Feld 330
D-69120 Heidelberg
Germany.
Tel: 49-6221-548913, 49-6221-808439 Fax: 544998
E-mail: js3@ix.urz.uni-heidelberg.de

Specialisation : Modern History.

Areas of Current Research:
South Asian Strategic Studies; International Relations of South Asian States; Modern History of South Asia.

Main Work at Hand:
Indo-Pakistan Crisis of 1990; Kashmir Dispute.

Recent/Forthcoming Publications:
Arms Procurement Policy of Pakistan (SIPRI - Sweden); *The Politics of Boundary Awards* (History Commission of Pakistan - Pakistan); Economic Trends, National Security & Defence Spending in Pakistan (RCSS, Sri Lanka).

GHAYUR, Sabur (M/42)
F.E.S., 3-B, Street 10, F-8/3
Islamabad, Pakistan.
Tel: 92-51-254112(W) 282088(H)
Fax: 92-51-260676(W)

Specialisation : Economics.

Areas of Current Research:
Employment & Manpower Development; Industrial Relations, Working Conditions of Unionization; Regional Cooperation.

Main Work at Hand:
Democracy & Trade Unions in Pakistan; Child Labour.

Recent/Forthcoming Publications:
Employment Generation & Poverty Alleviation in South Asia, ed. (Friedrich Ebert Stiftung, Islamabad, 1995); *Development, Governance & Governability*, (FES 1995).

HASAN, Hidayat (M/34)
Hagler Bailly Pakistan (pvt.) Ltd., 1, Street 15, F-7/2
Islamabad 44000, Pakistan.
Tel: 92-51-276113-8(W)
Fax: 92-51-824484(W)

Specialisation : Physics.

Areas of Current Research:
Command & Control Systems of Nuclear Weapons; Environmental Impacts of Nuclear Weapon Production, Use & Testing.

Main Work at Hand:
Command & Control of Nuclear Weapons in South Asia; Air Quality Impact of Uranium Mining.

HASSAN, Ibne (M/57)
18, Street 47, F-7/1
Islamabad, Pakistan.
Tel: 92-51-8240-38(W)
Fax: 92-51-2127-86(W)

Specialisation : International Relations; Law.

Areas of Current Research:
Regionalism & Globalism; Indian Ocean & Regional Security; Kashmir Issue & Regional Security.

Main Work at Hand:
Pakistan's Foreign Policy; Pakistan's Defence Policy.

Recent/Forthcoming Publications:
Effects of Great Power Veto on UN; Free Zones & World Court; Amnesty International as Human Rights Organization.

HILALY, Shameem (F/48)
House No.10A, Street 28 F6/1
Islamabad, Pakistan.
Tel: 91-51-821316(W)
Fax: 91-51-815414(W)

Specialisation : English Literature.

Areas of Current Research:
Women in Agriculture: Related Issues; Gender Discrimination in Govt. School Text Books; Supplementary Reading Material for New Literates.

Main Work at Hand:
Radio Programme on Reproductive Health; Radio Project on Women's Empowerment Through Agricultural Technologies.

HOODBHOY, Pervez (M/45)
Physics Dept., Quaid-i-Azam University
Islamabad, Pakistan.
Tel: 92-51-213429(W)

Specialisation : Physics.

Areas of Current Research:
Nuclear Proliferation.

HUSSAIN, Khadim (M/47)
House 492, Street 47, G-10/4
Islamabad, Pakistan.

Areas of Current Research:
Politics; Ethnicity; Security.

Recent/Forthcoming Publications:
"Jharkhand Movements in Perspective", *Regional Studies,* Vol.12, No.4;
"Caste System & Indian Policy", *Regional Studies,* Vol.12, No.2, Spring
1994; "Elections 1991", *Regional Studies,* Vol.10, No.2, Spring 1992.

HUSSAIN, Ijaz (M)
Dept. of International Relations, Quaid-e-Azam University
Islamabad, Pakistan.

Specialisation : International Law.

Areas of Current Research:
Kashmir Dispute; International Law; South Asian Security & Foreign Policies.

Recent/Forthcoming Publications:
Kashmir Dispute: An International Law Perspective; Independent Kashmir,
A Political Analysis.

INAYATULLAH, (M/64)
3, Street 18, F-7/2
Islamabad, Pakistan.

Specialisation : Political Science.

Areas of Current Research:
Ethnicity; Military's Role in Politics; Development in Social Sciences in
Pakistan.

Main Work at Hand:
State Security & Democracy in Pakistan; Social Sciences in Pakistan.

IRFANI, Surosh (M/47)
Institute of Strategic Studies, F-5/2, P.O. Box 1173
Islamabad, Pakistan.
Tel: 92-51-9204423(W)

Specialisation : Psychology & Education.

Areas of Current Research:
Culture & International Relations; Politics of Identity; Kashmir.

Main Work at Hand:
Modernity & New Discources in Iran; Feminism, Mysticism & Modernity.

Recent/Forthcoming Publications:
Contributions to *American Journal of Islamic Social Science,* Spring 1996.

JAFAR, Ghani (M/43)
>Institute of Strategic Studies, F-5/2, P.O. Box 1173
>Islamabad, Pakistan.
>Tel: 92-51-9204423(W) 256931(H)
>Fax: 92-51-9204658(W)
>
>**Specialisation** : Law.
>
>**Areas of Current Research**:
>South Asia; West Asia.
>
>**Main Work at Hand**:
>China & the Muslim Third World; The Brahamanic Bomb.
>
>**Recent/Forthcoming Publications**:
>"The West Vs Muslims, The Moments of Truth", *Strategic Perspectives,*
>Vol.3, No.1&2, Winter 94 & Spring 95; "Kashmir & Palestine, Similar
>Solution to Similar Issues", *Strategic Studies,* Vol. 18, No 243, Winter 95,
>Spring 96.

KAZI, Shanaz (F/46)
>Pakistan Institute of Development Economics
>Quaid-e-Azan University, Islamabad, Pakistan.
>Tel: 92-51-216947(W)
>
>**Specialisation** : Economics.
>
>**Areas of Current Research**:
>Employment; International Migration; Women & Development.
>
>**Main Work at Hand**:
>"Some Measures of the Status of Women in the Case of Development in
>South Asia", *Women & Development, South Asia,* (Macmillan New Delhi);
>"Re-integration of Return Migrants - The Asian Experience", *Pakistan
>Development Review,* 1994.

KHALID, Rasheed (M)
>Dept. of Defence & Strategic Studies, Quaid-e-Azam University,
>Islamabad, Pakistan.
>
>**Specialisation** : International Relations.
>
>**Areas of Current Research**:
>Disarmament & Arms Control.

KHAN, Amir (M/25)
>House No. 31, Street No.12
>F-6/3, Islamabad, Pakistan.
>
>**Specialisation** : International Relations.

Areas of Current Research:
South Asia; Peace Studies; Nuclear Debate in South Asia.

Main Work at Hand:
US Role in South Asia in Post Cold War Era.

KHAN, Ashfaque Hasan (M/42)
Chief of Research, Pakistan Institute of Development Economics,
P.O. Box 1091, Islamabad, Pakistan.
Tel: 92-51-824574(W)

Specialisation : Economics.

Areas of Current Research:
Macro/Monetary/International Economics; Public Finance; Regional Co-operation.

Main Work at Hand:
Uruguay Round & it's Implications for Pakistan; Financial Sector Reforms &
Monetary Policy in Pakistan.

Recent/Forthcoming Publications:
"Financial Liberaliazation, Savings & Economic Development", in *Economic
Development & Cultural Change*; "Trade Strategies to the Post Uruguay
Round Trading Environment: A Case of Pakistan", UNCTAD; *Adoption of
Trade Strategies*, (UN, New York).

KHAN, Aurangzeb Zulfiqar (M/30)
Institute of Strategic Studies, F-5/2, P.O. Box 1173
Islamabad, Pakistan.
Tel: 92-51-218406/7(W) 260727(H)
Fax: 92-51-218407(W)

Specialisation : Buisness Management.

Areas of Current Research:
Economic Co-operation Between States of South Asia; Contemporary
Economic Development in the Muslim World.

Main Work at Hand:
"Prospects for Future Regional Economic Integration", *Strategic Studies,*
Vol.15, Winter 1992, No.2, pp.73-79; "South Asia: A Review & Appraisal of
Current Regional Economic Trends", *Strategic Perspectives,* Vol.2, Summer
1994, No.3, pp.77-95.

Recent/Forthcoming Publications:
India & Pakistan: Imperatives & Prospects for Regional Economic Co-operation; SAARC & ECO: Economic Policy Options for Pakistan.

KHAN, Muhammad Liyas (M/41)
Institute of Policy Studies, Nasr Chambers, Block 19
Markaz F-7, Islamabad, Pakistan.
Tel: 92-51-818230(W) 05261-2916(H)
Fax: 92-51-824704(W)

Specialisation : Central Asian Studies.

Areas of Current Research:
Russian Politics/Society; Central Asian Affairs; South Asia - Russia/Central
Asia Relations.

Main Work at Hand:
Muslim Minorities in Russian Federation & Non-Muslim CIS Countries;
Islamic Revivalism in Central Asia: A Case Study of Tajikistan.

Recent/Forthcoming Publications:
Civil War in Tajikistan: Origins, Evolution & Prospects for Peace, (May
1995); Kazakhstan: Averting Political Instability or Reverting Once Again to
Soviet Era Depositsm, (March 1995).

KHAN, Zaffarullah (M/30)
President, Green Press
G.P.O.Box 1123
Islamabad 44000, Pakistan.
Tel: 92-51-823685, 819840(W)
Fax: 92-51-271231(W)
E-mail: zafar@press.sdnpk.undp.org

Specialisation : International Relations, Politics.

Areas of Current Research:
Role of Media; People's to People's Democracy (NGO's Role as Well);
South Asian Politics.

Recent/Forthcoming Publications:
Facts & Fiction About NGO's; State of Media in Pakistan.

KHATTAK, Saba Gul (F/34)
Sustainable Development Policy Institute
3, UN Bulevard, Diplomatic Enclave
Islamabad, Pakistan.
Tel: 92-51-277146(W)
Fax: 92-51-278135(W)
E-mail: saba@sdpi.sdnpk.undp.org

Specialisation : Political Science.

Areas of Current Research:
Comparative Politics, Political Theory Including Feminist Theory, State Theory.

Main Work at Hand:
Individual Policy in Pakistan; Alternatives on Security & the State in Pakistan.

Recent/Forthcoming Publications:
Women's Movement & the State in Pakistan, Women Refugees in Pakistan.

KORESHI, S.M. (M/69)
Institute of Policy Studies, Nasr Chambers, Block 19
Markaz F-7, Islamabad, Pakistan.
Tel: 92-51-818230(W)
Fax: 92-51-824704(W)

Specialisation : Political Science.

Areas of Current Research:
Middle East; Muslim World.

Recent/Forthcoming Publications:
Contemporary Power Politics & Pakistan: Western Fundamentalism in Action.

MAHMOOD, Mir Annice (M/45)
Pakistan Institute of Development Economics, P.O. Box 1091
Islamabad, Pakistan.
Tel: 824070-80(W)
Fax: 92-51-210886(W)

Specialisation : Economics.

Areas of Current Research:
Education; Health; Trade & Politics.

Main Work at Hand:
Transport in South Asia.

MAHMOOD, Moazam (M/41)
Chief of Research, Pakistan Institute for Development Economics,
P.O. Box 1091, Islamabad, Pakistan.
E-mail: arshad@pide.sdnple.undp.org

Specialisation : Economics.

Areas of Current Research:
Macro Management; Growth & Distribution; Micro Production & Consumption Behaviour.

Main Work at Hand:
Recent Determinants of Inflation in Pakistan; Child Labour, Fertility & the Labour Market.

Recent/Forthcoming Publications:
Just Adjustment (ed.) Banuri T. & Mahmuud M., (OUP); *Capital Accumulation in Agriculture*, (OUP).

MAHMOOD, Naushin (F/45)
PIDE, P.O. Box 1091
Quaid-e-Azam University, Islamabad 44000, Pakistan.
Tel: 92-51-824070(W)
Fax: 92-51-210886(W)

Specialisation : Sociology.

Areas of Current Research:
Demographic Change in Pakistan & Asian Countries; Population & Health; Gender Issues & Household Dynamics.

Main Work at Hand:
Desired Fertility in Pakistan: Influence of Husbands in Decision Making; Gender Differences in Reproductive Goals in Pakistan: Implications for Fertility Outcomes.

Recent/Forthcoming Publications:
Motivation for Fertility Control Behaviour in Pakistan; Health Care Determinants of Child Mortality in Pakistan.

MAHMOOD, Zafar (M/42)
Pakistan Institute of Development Economics, P.O. Box 1091
Islamabad, Pakistan.
Tel: 824070(W)
Fax: 92-51-210886(W)
E-mail: arshad%pide@sdnpk.undp.org

Specialisation : Economics.

Areas of Current Research:
International Economics; Industrial Economics; Labour Economics.

Main Work at Hand:
Implication of the Uruguay Round for Pakistan; Prospects of Pak-India Trade.

Recent/Forthcoming Publications:
"Potential of Small & Medium-Sized Industries in Pakistan", (ESCAP); *Emerging Global Trading Environment & Pakistan*, (Oxford Press); "Determinants of Misinvoicing of Trade", *Journal of International Development*.

MAHMUD, Ershad (M/25)
Institute of Policy Studies, Nasr Chambers, Block 19
Markaz F-7, Islamabad, Pakistan.
Tel: 92-51-818230(W)
Fax: 92-51-824704(W)

Specialisation : Political Science.

Areas of Current Research:
Kashmir Issue; Pak-India Relations.

Recent/Forthcoming Publications:
Options on Kashmir; Siachen Issue.

MAHMUD, Khalid (M/60)
Editor-in charge, The Nation, Zero Point
Islamabad, Pakistan.
Tel: 92-51-278355(W) 251539(H)
Fax: 92-51-278353(W)

Specialisation : Political Science.

Areas of Current Research:
Political Parties & Process in the Sub Continent.

Recent/Forthcoming Publications:
The Indian Political Scene, (Published by the Institute of Regional Studies,
Islamabad, 1989).

MASUD, Salma (F/25)
8, Baazar Road, G-6/4
Islamabad, Pakistan.
Tel: 92-51-812927(W)
Fax: 92-51-221256(W)
E-mail: salma@masuds.sdnpk.undp.org

Specialisation : Economics.

Areas of Current Research:
Gender & Development in South Asia; Child Nutrition & Development;
Employment Opportunities in SAARC Countries.

Main Work at Hand:
Gender Comparison in South Asia; Legal Discrimination Against Women in
Pakistan.

Recent/Forthcoming Publications:
Contribution to Human Development Report for South Asia.

MATEEN, Amir (M/29)
Chief Reporter
The Nation, Nawa-i-Waqt Building,
Zero Point, Islamabad, Pakistan.
Tel: 92-51-824006(W)

Specialisation : English Law.

Areas of Current Research:
Politics (South Asian Domestic); Environment.

Main Work at Hand:
Reporting for *The Nation*, (An English Daily).

MAZARI, Shireen M. (F/44)
Editor, Weekly Pulse, Office No.2, Block 46
Jinnah Avenue, Blue Area, Islamabad, Pakistan.
Tel: 92-51-825497(W) 92-51-828201(H)
Fax: 92-51-825497(W)

Specialisation : Security Studies.

Areas of Current Research:
Nuclear Proliferation; International Trade; Weapons Proliferation.

MIAN, Zia (M)
Sustainable Development Policy Institute
P.O. Box 2342,
Islamabad, Pakistan.

Specialisation : Physics.

Areas of Current Research:
South Asia, Peace, Nuclear Power, Weapons.

MOHIB, Ul Haq Sahibzada (M/57)
Senior Research fellow, Institute of Policy Studies
Nasr Chambers, Block 19,Markaz F-7, Islamabad, Pakistan.
Tel: 92-51-818230(W)
Fax: 92-51-824704(W)

Specialisation : Agriculture.

Areas of Current Research:
Agriculture; Rural Development; Economics.

Main Work at Hand:
Rural Credit, Pakistan; Rural Employment, Pakistan; Review of Policies &
Programmes.

Recent/Forthcoming Publications:
Pakistan: Population & Food, (IPS, 1993); *Critical Issues in Agriculture* (ed.), (IPS, 1995); *Emerging Role of Stock Markets in Pakistan Economy* (ed.), IPS, 1995

NAIK, Niaz A. (M/69)
Pakistan Security & Development Association
P.O. Box 2306, Islamabad, Pakistan.
Tel: 92-51-823730(H)

Specialisation : History.

Areas of Current Research:
Peace & Security; Arms Control & Disarmament; Sustainable Development.

NAQVI, Nauman (M/27)
No.46, Street 12, F-6/3
Islamabad, Pakistan.
Tel: 92-51-278134/278136(W)
Fax: 92-51-278135(W)
E-mail: nauman%sdpi@sdnpk.undp.org

Specialisation : Philosophy.

Areas of Current Research:
Critique of Development; Modernity/Postmodernity Debates; Effects of Globalisation, Especially in South Asia.

Main Work at Hand:
Rethinking Security, Rethinking Development, 1996, An Anthology of Papers from the 3rd Annual South Asian NGO Summit; Green Economics.

NASEEM, Muhammad Ayaz (M/32)
Dept. of Defence & Strategic Studies
Quaid-i-Azam University
Islamabad, Pakistan.
Tel: 92-51-828616(W)
Fax: 92-51-821397(W)
E-mail: ayaz%dss.gav@sdnpk.undp.org

Specialisation : International Relations.

Areas of Current Research:
Political Economy; International Relations (Foreign Policy); Environment & Security.

Main Work at Hand:
Culture of Peace in South Asia; Environment & Security.

Recent/Forthcoming Publications:
Development of Culture of Peace in South Asia: The Case of Pakistan (UNESCO, 1996); *Rising Militancy & Ethnic Violence in South Asia* (1996).

NAYYAR, Abdul Hameed (M/50)
Dept. of Physics, Quaid-i-Azam University
Islamabad 45320, Pakistan.

Specialisation : Physics.

Areas of Current Research:
Nuclear issue in South Asia.

Main Work at Hand:
"Fissile Material Production Potential in South Asia", *Science & Global Security*; The Clock is Ticking for South Asia, *Bulletin of the Atomic Scientists*.

QAZI, Najeeb Rehman (M)
Research Fellow, Institute of Policy Studies
Nasr Chambers, Block 19, Markaz F-7, Islamabad, Pakistan.
Tel: 92-51-818230(W) 2922419(H)
Fax: 92-51-824704(W)

Specialisation : Economics.

Areas of Current Research:
Public Finance; Labour Economics; Economics & Education.

Main Work at Hand:
Fifty Years of Pakistan Economy; Impact of Education on Income Distribution.

Recent/Forthcoming Publications:
"Earning Profiles of Dept. Heads" (Coauthor) *Industrial Labour Relations Review*, Jan 1996; "Export Earnings Instability in Pakistan", *Pakistan Development Review*.

QUTUB, Syed Ayub (M/46)
Executive Director
Pakistan Institute for Environment
Development Action Research (PIEDAR)
Second Floor
Yasin Plaza, 74-w, Blue Area
Islamabad, Pakistan.
Tel: 92-51-820454(W) 92-51-9203972(H)
Fax: 92-51-276507(W)

Specialisation : Economic Geography.

Areas of Current Research:
Environmental Management; Sustainable Development; Community Empowerment.

Main Work at Hand:
Community based Irrigation Management in the Wheat-Cotton Belt of Irrigated Punjab, Pakistan.

Recent/Forthcoming Publications:
"Environmental Impact of Agricultural Policies", *Agricultural Policy Analysis,* with W.V. der Geest, (OUP).

RAFIQUE, Syed Nazam (M/34)
Institute of Strategic Studies, F-5/2
Islamabad, Pakistan.

Specialisation : Defence & Strategic Studies.

Areas of Current Research:
US Policy in South Asia; US/Europe in the Post Cold War Era; Nuclear Proliferation.

Recent/Forthcoming Publications:
Europe: Struggling for a Common Defence Policy; Examining Asia's Nuclear Rivalries: China-India & India-Pakistan.

RAIS, Rasul Bakhsh (M/47)
Director, Area Study Centre
Quaid-i-Azam University, Islamabad 45320, Pakistan.
Tel: 92-51-825809(W)
Fax: 92-51-821397(W)
E-mail: rbrais@asc-qau.sdnpk.undp.org

Specialisation : Political Science.

Areas of Current Research:
Indian Ocean; South Asian Security; Democratic Transition in Pakistan.

Main Work at Hand:
Pakistan's Search for Security; Pakistan's Transition to Democratic Rule.

Recent/Forthcoming Publications:
State, Society & Democratic Change in Pakistan (Edited Volume, Oxford University Press, Karachi); *War Without Winners: Afghanistan's Uncertain Transition After the Cold War,* Co Ed. with Charles Kennedy, (OUP 1994).

SABOOR, Ali Syed (M/24)
Institute of Policy Studies, Farlaiz F/7, Nasr Chamber
Markaz F-7, Islamabad, Pakistan.

Areas of Current Research:
Kashmir Issue; Pakistan-India Relations; American Interest in South Asia.

Main Work at Hand:
Plight of the Indian Muslims; Myths & Realities India-Bangladesh Relations.

SAEED, Amera (F/52)
Institute of Regional Studies, NAFDEC Complex, 56-F Blue Area
F-6/1, Islamabad 44000, Pakistan.
Tel: 92-51-9204934(W)
Fax: 92-51-9204055(W)

Specialisation : English Literature/Political Science.

Areas of Current Research:
Indian Ocean Region & South Asia; Asia-Pacific Region & Regional Organizations; Local Self Governing Institutions.

Main Work at Hand:
Pakistan's First Ten Years of Foreign Policy & Defence Formulation; Bangladesh Politics - Focus on the Latest Elections.

Recent/Forthcoming Publications:
BJP at the Cross Roads; An Economic Forum for the Indian Ocean Region; "Indian State Elections", (Published by Institute of Regional Studies, Islamabad).

SAIYID, Dushka Hyder (F/45)
Dept. of History, Quaid-i-Azam University
Islamabad, Pakistan.

Specialisation : South Asian History.

Areas of Current Research:
Punjab in the British Period; Muslim Women of South Asia; Women in Islam.

Main Work at Hand:
Exporting Communism to India; Why Moscow Failed; Islamabad: National Commission of Historical & Cultural Research, 1995.

Recent/Forthcoming Publications:
Changing Position of Muslim Women in Punjab, 1872-1947, (Macmillan UK).

SAMEE, Mehreen (F/31)
Sustainable Development Policy Institute
No. 46, Street 12, F-6/3
Islamabad, Pakistan.

Tel: 92-51-278134(W)
Fax: 92-51-278135(W)
E-mail: mehreen@sdpi.sdnpk.undp.org (W)

Specialisation : Economics & Politics of Development.

Areas of Current Research:
Trade & Environment.

Main Work at Hand:
Liberalisation & it's Impact on Trade in Cotton: A Cross Regional Study.

SATHAR, Zeba (F/42)
The Population Council, No. 55, Street 1, F-6/3
Islamabad, Pakistan.
Tel: 92-51-277439(W)

Specialisation : Democracy.

Areas of Current Research:
Fertility, Family Planning, Women's Status.

Main Work at Hand:
Situation Analysis of VBFPW; Women's Status & Fertility in Panjab.

SATTAR, Abdul (M/64)
7, College Road, F-7/3, Islamabad
Pakistan.

Specialisation : Geography; International Law.

Areas of Current Research:
Nuclear Issue in South Asia; Pakistan's Foreign Policy.

Recent/Forthcoming Publications:
"Reducing Nuclear Dangers in South Asia", *The Non-Proliferation Review*,
(California, USA., Winter 1994; "Simla Agreement - Negotiation Under
Duress", *Regional Studies*, 1995.

SAYOOD, Syed Abu (M/70)
Institute of Policy Studies, Nasr Chambers, Block 19
Markaz F-7, Islamabad, Pakistan.
Tel: 92-51-818230(W) 271008(H)
Fax: 92-51-824704(W)

Specialisation : Literature.

Areas of Current Research:
Pakistan Political Parties/System; Geo-Strategic Studies & Implication of
South Asian States with Specific References to Pakistan; Ethnic/Communal/
National Movements in the Region.

SHAH, Aqil (M/23)
783, Street 64,, G-9/4
Islamabad, Pakistan.

Specialisation : International Relations.

Areas of Current Research:
Indian Politics & Foreign Policy; CBMs; Civil Military Relations.

Main Work at Hand:
Confidence - Building in South Asia Through People-to-People Contacts: A Pakistani Perspective.

SIDDIQUI, Nadeem (M/30)
Lecturer in Modern History, Dept. of History
Quaid-i-Azam University, Islamabad, Pakistan.

Specialisation : US Studies.

Areas of Current Research:
Modern South Asian History; Conflict Resolution; US Policy Towards South Asia.

Main Work at Hand:
Muslim Political Thought in South Asia; Conflict Resolution & Environmental Aspect of International Politics.

Recent/Forthcoming Publications:
"Comparing Third Parties in Peace Processes: Afghanistan, Guatemala & Sri Lanka", in Goran Lindgren (ed.), *Experiences from Conflict Resolution in the Third World*, (Uppsala, 1993); "Muslim Political Thought in South Asia", *National Development and Security*, Vol. II, No. 4, May 1995.

SYED, Shamoon Hashmi (M/25)
House No. 273, Street 75
G-9/3, Islamabad, Pakistan.

Specialisation : Political Science.

Areas of Current Research:
Politics of South Asia; Current Affairs; Art & Culture.

TAHIR, Amin (M/44)
Institute of Policy Studies, Nasr Chambers, Block 19
Markaz F-7, Islamabad 44000, Pakistan.
Tel: 92-51-818230(W)
Fax: 92-51-824704(W)

Specialisation : Political Science.

Areas of Current Research:
International Relations; Politics of South West Asia, South Asia, Central Asia; Pakistani Society & Ethnicity.

Main Work at Hand:
Mass Resistance in Kashmir; Ethno-National Movements in the Sindh Province of Pakistan.

Recent/Forthcoming Publications:
Mass Resistance in Kashmir, (IPS, 1995).

TAHIR, Rizwan (M/33)
School of Economics, IIIE
P.O. Box 1687, IIU, Islamabad, Pakistan.

Specialisation : Economics.

Areas of Current Research:
Islamic Economics; Applied Econometrics; Econometric Modelling.

Main Work at Hand:
Defence & Economic Growth: Case of Some Selected LDCs.

Recent/Forthcoming Publications:
Islamisation of Economy: A Case Study of Pakistan; Defence Spending & Economic Growth: Re-examining the Issue of Casualty for India & Pakistan.

USMANI, Maheen (F/27)
Human Development Centre, 8 Bazar Road
G-6/4, Islamabad, Pakistan.

Specialisation : Political Science.

Areas of Current Research:
Political Institutions of South Asia.

ZEHRA, Nasim (F)
1st Floor, House No.75, Street No.1, F-6/3
Islamabad, Pakistan.

Specialisation : International Affairs.

Areas of Current Research:
National Security; Pakistani Politics; South/South West Asia.

ZIA, Shehla (F/48)
Aurat Foundation, House 10
Street 28, F-6/1, Islamabad, Pakistan.
Tel: 92-51-821316(W)
Fax: 92-51-815414(W)

Specialisation : Law.

Areas of Current Research:
Women's Issues in Law & Politics.

Main Work at Hand:
Discriminatory Legislation Against Women.

Recent/Forthcoming Publications:
Legislative Watch.

ZIAUDDIN, Mohammad (M/56)
Dawn, Zero Point
Islamabad, Pakistan

Specialisation : Journalism.

■ *Jamshoro*

CHACHAR, Abdul Khalique (M/52)
Associate Professor & Chairman, Dept. of International Relations
University of Sindh, Jamshoro, Pakistan.
Tel: 91-221-771681-90(W) 91-221-651529(H)

Specialisation : Strategic Studies.

Areas of Current Research:
Nuclear Proliferation in South Asia; International Political Economy; Pakistan's Security Concerns & South Asia.

Recent/Forthcoming Publications:
The End of Cold War: Historical Perspectives, 1994 ; The case of Proliferation of Nuclear Weapons in South Asia, 1992; From I.T.O. to W.T.O.: The Changing International Trade Regime & Pakistan.

SHAH, Methab Ali (M/44)
Dept. of International Relations, Sindh University
Jamshoro, Sindh, Pakistan.
Tel: 91-42-221-771575(H)

Specialisation : International Relations.

Areas of Current Research:
Theory of International Relations; Ethnicity & Geo-Politics.

Main Work at Hand:
Ethnic Conflicts in Pakistan & their Implications for National Security.

Recent/Forthcoming Publications:
Ethnicity in Foreign Policy of Pakistan: 1971-1994.

■ *Karachi*

ABBAS, Azhar (M/62)
Monthly Herald, Haroon House
Dr. Ziauddin Ahmed Road, Karachi 74200, Pakistan.
Tel: 92-21-569-9168(W)
Fax: 92-21-5689260(W)

Specialisation : Economics.

Areas of Current Research:
Media & Emerging Democracies; Environment; Ethnic & Regional Conflicts.

Recent/Forthcoming Publications:
Army, People and Democracy; Media in a Peace Society.

AHMAR, Moonis (M/36)
Dept. of International Relations, University of Karachi
Karachi 75270, Pakistan.
Tel: 92-21-479001(W) 92-21-476356(H)
Fax: 92-21-4963373(W)

Specialisation : European Studies.

Areas of Current Research:
Confidence Building Measures; Conflict Management & Resolution;
South & Central Asian Studies.

Main Work at Hand:
Contemporary Central Asia; Erosion of State Power in Pakistan: A Study of
Karachi Crisis; Pakistan & Israel - Distant Enemies.

Recent/Forthcoming Publications:
Peace Process: Lessons for India & Pakistan from the Middle East; Ethnic
Conflicts in Central Asia: Pakistan's Perspective.

ALI, Marvi Memon (F/23)
Citibank N.A., Bahria Complex
4th Floor, M.T. Khan Road, Karachi, Pakistan.
Tel: 92-21-5611539(W)
Fax: 92-21-5615342(W)

Specialisation : International Relations

Areas of Current Research:
Self Determination/Nationalism; Systemic Theories of International Relations; Morality/Ethics in International Relations.

Main Work at Hand:
"The Soviet Leadership in Tightrope", *Pakistan Horizon*.

Recent/Forthcoming Publications:
Self Determination & the UN.

ASKARI, Mirza Hasan (M/70)
c/o Dawn, Haroon House
Ziauddin Ahmed Road, Karachi, Pakistan.
Tel: 92-21-111444777(W) 5678803(H)
Fax: 5682187(W)

Specialisation : English Literature.

Areas of Current Research:
International Relations; South Asian Affairs.

GHAUS, Khalida (F/37)
Dept. of International Relations, University of Karachi
Karachi, Pakistan.

Spécialisation : International Relations.

Areas of Current Research:
Conflict Resolution & Management; Human Rights; Foreign Policy.

Main Work at Hand:
Conflict Resolution: A Theoretical Frame Work; Peace in Asia.

Recent/Forthcoming Publications:
UN & Human Rights; UN & Crisis Management; The Impact of Social
Revolutions on the Making of Human Rights.

HASAN, Sabiha (F/42)
Assistant Professor, Pakistan Study Centre
University of Karachi, Karachi, Pakistan.
Tel: 92-21-962497(W), 5833033(H)

Specialisation : International Relations.

Areas of Current Research:
Foreign Policy of Pakistan with Special Reference to the Role of the Military; Alternate Development Theories in the Post Cold War Period, State
System in the Post Cold War Era.

Main Work at Hand:
Editing *Pakistan Perspectives* (PSC Research Journal) Directing Students
Community Research Action Programme (SCRAP) at PSC, Editing SCRAP
Newsletters.

HASAN, Rubab (F/45)
Area Study Centre for Europe, Univeristy of Karachi
Karachi, Pakistan.

Specialisation : International Relations.

Areas of Current Research:
South Asia; Europe.

Main Work at Hand:
A Chronology of Events in Eastern Europe & Soviet Union - 1989.

HUSSAIN, Sahid (M/48)
Senior Editor, Newsline Magazine
D-6-Krekashan, Block-9, Clifton, Karachi, Pakistan.

Specialisation : Physics.

Areas of Current Research:
Nuclear Issue; Regional Politics; Pakistan's Domacratic Politics.

KHAN, Fatehyab Ali (M/60)
Chairman, Pakistan Institute of International Affairs
Aiwan-e-Sadar Road,
P.O. Box 1447, Karachi 74200, Pakistan.

Specialisation : Political Science.

Areas of Current Research:
Political Science & International Relations.

Main Work at Hand:
Supreme Court Judgement (Judges Case 1996).

Recent/Forthcoming Publications:
Various Articles on Constitutional Law in *Dawn*, *The Muslim* & *Pakistan Horizon*.

KHAN, Hina (F/21)
Lecturer, Dept. of General History
University of Karachi, Karachi 75270, Pakistan.
Tel: 92-21-6331931(H)
Fax: 92-91-8116704(H)

Specialisation : General History.

Areas of Current Research:
Peace & Conflict Studies with Particular References to South Asia; Peace Keeping - Both UN & Non UN; Government & Politics of Central Asia.

Main Work at Hand:
Peace Keeping - A Conceptual Study; Non-Military Sources of Conflict in South Asia.

Recent/Forthcoming Publications:
"The Conquest of Central Asia: The Tsarist Phase", *Pakistan Horizon*; "ECO - A Futuristic Study", *Third World*, May 1995.

MAHMOOD, Tehmina (F/31)
Research Officer, Pakistan Institute of International Affairs
Aiwan-e-Sadar Road, P.O. Box 1447,
Karachi 74200, Pakistan.

Specialisation : International Relations.

Areas of Current Research:
South Asia; Central Asia.

Main Work at Hand:
Editing & Publication of *Pakistan Horizon* Research Work.

Recent/Forthcoming Publications:
UN & Nuclear Issue in South Asia; Pakistan's Foreign Policy in Post Cold War World.

MEHDI, Syed Sikander (M/47)
Professor, Dept. of International Relations
University of Karachi, Karachi 75270, Pakistan.
Tel: 92-21-8116705(H)

Specialisation : International Relations.

Areas of Current Research:
Peace & Security Issues with Particular References to South Asia; UN & Other Peace Keepings; Refugee & Migration Issues.

Main Work at Hand:
Changing Security Discourse in Pakistan; Hazards of Peace Keeping.

Recent/Forthcoming Publications:
"Co-operative Security in Europe & South Asia", "Towards a New Security Discourse in Pakistan", *National Development & Security,* Vol.4, No.3, Feb 1996; "European Union & Asian Future", *Third World*, Vol.20, No.4, April 1996.

MOHIUDDIN, Lubna (F/25)
Pakistan Institute of International Affairs
Aiwan-e-Sadar Road,
P.O. Box 1447, Karachi 74200, Pakistan.

Specialisation : International Relations.

Areas of Current Research:
South Asia; Human Rights.

NAQVI, M.B. (M/67)
B-116 Block I, North Nazimabad
Karachi 74700, Pakistan.

Specialisation : Economics.

Areas of Current Research:
Politics - National Security Questions; Nuclear Weapons - International
Relations; Economics & International Economics.

Main Work at Hand:
Columnist of *Daily Dawn*, Karachi.

Recent/Forthcoming Publications:
Mohajirs of Pakistan; Nuclear Agony of South Asia; Pakistan's Foreign
Policy.

QURESHI, Yasmin (F/42)
Oxford University Press, 5, Bangalore Town
Sharae Faizal, Karachi, Pakistan.
Tel: 92-21-4529025-8(W) 5861196(H)
Fax: 92-21-4547640(W)
E-mail: oup@oup.khi.erum.com.pk

Specialisation : History.

Areas of Current Research:
The Problem of Ethnicity & Integration in Pakistan; Kashmir.

Main Work at Hand:
The Destruction of Pakistan's Democracy; Management of Third World
Crisis: Theory & Practice.

SHAH, Syed Imdad (M/45)
Area Study Centre for Europe, University of Karachi
University Road, Karachi, Pakistan.
Tel: 92-21-4973610(W) 448647(H)

Specialisation : International Relations.

Areas of Current Research:
Pakistan-European Relations; Pakistan-Turkish Relations; Central European /
Eastern European Topics.

Main Work at Hand:
Journal of European Studies; The British Political Scene: The Re-emergence of the Labour Party.

Recent/Forthcoming Publications:
"Pakistan-German Ties 1985-95", *Journal of European Studies,* Vol.11 & 12, No.2 & 1, July 1995 & Jan. 1996; The Kashmir Problem: British Role in Historical Perspective.

SHAKOOR, Farzana (F)
Senior Research Officer, Pakistan Institute of International Affairs
Aiwan-e-Sadar Road
P.O.Box 1447, Karachi 74200, Pakistan.

Specialisation : Pakistan Studies.

Areas of Current Research:
South Asian Studies.

Main Work at Hand:
Editing & Publication of *Pakistan Horizon* & Research Work.

Recent/Forthcoming Publications:
UN & Kashmir; South Asia: Foreign Policy Trends in the Post Cold War Era; "Bhutan: The Issue of Ethnic Divide", *Pakistan Horizon,* Vol.48, April 1995.

SYED, Humayun (M/45)
Professor, Dept. of Political Science
University of Karachi, Karachi, Pakistan.

Specialisation : Political Science.

Areas of Current Research:
Politics of Ethnicity; Local Government with Special References to Pakistan.

Main Work at Hand:
A Select Bibliography: Local Government & Administration in Pakistan.

Recent/Forthcoming Publications:
Sheikh Mujibur Rahman's, 6- Point Formula: An Anlytical Study of Breakup of Pakistan, (Royal Book Company, 1995).

TAHA, Syed Mohammed (M/30)
Lecturer, Dept. of General History
University of Karachi, Karachi 75270, Pakistan.

Specialisation : History/International Relations.

Areas of Current Research:
Middle East & South Asia.

Main Work at Hand:
US & Peace Building in South Asia: A Historical Study.

Recent/Forthcoming Publications:
Conflict Resolution: An Analysis; Religion as a Source of Conflict in India,
(Hamdard Islamians).

TAHIR, Naveed Ahmad (F/43)
Professor/Acting Director, Area Study Centre for Europe
University of Karachi, Karachi, Pakistan.

Specialisation : European Studies.

Areas of Current Research:
Europe & South Asia.

Main Work at Hand:
Germany in Post Cold War Europe; Post Cold War European Order & South
Asia.

Recent/Forthcoming Publications:
The Expansions of the European Areas: Problems & Prospects, 1995.

WIZARAT, Talat Ayesha (F/49)
Chairperson, Dept. of International Relations
University of Karachi, Karachi, Pakistan.

Specialisation : International Relations.

Areas of Current Research:
Middle East; South Asia; Conflict Resolution.

Main Work at Hand:
Strategies for Peace & Security in the Persian Gulf; Political Framework of
Conflict & Co-operation in South Asia.

Recent/Forthcoming Publications:
CIS & ECO: Options for the Central Asian States in Contemporary Central
Asia.

■ *Lahore*

ADIL, Adnan (M/29)
The Friday Times, 45, The Mall
Lahore, Pakistan.
Tel: 92-42-7243779(W)
Fax: 92-42-7245097(W)

Specialisation : Political Science.

157

Areas of Current Research:
Internal Security; National Politics; World Trade; Foreign Policy Issue.

AGHA, Ayesha Siddiqa (F/29)
'Al-Sair', 31 Mira Khan Road
St. John's Park, Lahore Cantt., Pakistan.

Specialisation : War Studies.

Areas of Current Research:
Light Weapons; Arms Procurement; Defence, Decision Making & Management.

Recent/Forthcoming Publications:
Management of Light Weapons Manufacturing in Public & Private Sectors - A View from Pakistan; Self Sufficiency in Defence: The Collaboration Approach.

AHMED, Rafique (M)
Centre for South Asian Studies, Punjab University
Lahore, Pakistan.

Specialisation : Economics.

Areas of Current Research:
Pakistan - India Relations.

Main Work at Hand:
WTO & India - Pakistan Trade.

MALIK, Abdulla (M/75)
134, Tipo Block New, Garden Town
Lahore, Pakistan.

Areas of Current Research:
Indo-Pakistan Analytical History in the Context of South Asia;
Military's Role in 3rd World & Pakistan.

Main Work at Hand:
Biography of a Bureaucrat.

Recent/Forthcoming Publications:
Army & Pakistan; Basic Facts: Pakistan & The Armed Forces.

MIRZA, Sarfaraz Hussain (M)
House No.14, Abid Majied Road, Bazar No.28
Lahore, Pakistan.

Specialisation : Political Science.

Areas of Current Research:
Pakistani Movement; East-Pakistan Crisis - A Chronology (1947-71).

Main Work at Hand:
Intellectuals Role in the Struggle for Pakistan (Majlis-i-Kabir-Pakistan) 1939-1941.

MUMTAZ, Khawar (F/50)
208, Scotch Corner, Upper Mall
Lahore, Pakistan.
Tel:092-42-5760764(W)

Specialisation : International Relations.

Areas of Current Research:
Women in Development; Women in Sustainable Development.

QURAISHI, Munir Ashghar (M/22)
Dept. of Political Science, Punjab University
Lahore, Pakistan.
Tel:92-42-5863982(W)

Specialisation : Political Science.

Areas of Current Research:
Pakistan's Relations in the Region.

Main Work at Hand:
Pakistan-Afghan Relations: Post-Soviet Withdrawal Era.

RASHID, Ahmed (M/48)
108/10, Tofail Road, Lahore Cant
Pakistan.

Specialisation : Political Sience.

Areas of Current Research:
Afghanistan; Central Asia; Pakistan Foriegn Policy & Politics.

Main Work at Hand:
The Resurgence of Central Asia: Islam or Nationalism.

RASHID, Tehmina (F)
Dept. of Political Science, Punjab University,
Lahore, Pakistan.
Tel:92-42-5863982(W)

Specialisation : Political Science.

Areas of Current Research:
Pakistan Foriegn Policy; Pakistan's Political System.

REHMAN, I.A. (M/65)
13, 3rd Floor, Sharif Complex, Main Market,
Gulberg-II, Lahore 54660, Pakistan.
Tel:5750217(W) 7233913(H)
Fax:5713078(W)

Specialisation : History; Political Science.

Areas of Current Research:
Democratic Politics; Human Rights; Culture.

Main Work at Hand:
50 Years of Pakistan; Great Pakistanis (Series for Children).

Recent/Forthcoming Publications:
Arts & Crafts of Pakistan; *Pakistan Under Siege*, 1990.

SARWAR, Beena (F/32)
Editor, The News on Friday
13, Davies Road, Lahore 54000, Pakistan.
Tel: 92-42-6304745(W) 92-42-7234128(H)
Fax: 92-42-6371335(W)

Specialisation : Studio Art; Literature.

Areas of Current Research:
Human Rights; Women; Environment.

SETHI, Najam (M/49)
The Friday Times, 45 - The Mall
Lahore, Pakistan.
Tel: 92-42-7120781(W) 5710336(H)

Areas of Current Research:
Pakistan Politics & Economics

Main Work at Hand:
From Plunder to Blunder & Back - Pakistan Under Nawaz Sharif & Benazir Bhutto 1990-1996.

SOOFI, Ahmer Bilal (M/33)
Attorneys, Solicitors & Advocates, 13-Fane Road
Lahore, Pakistan.
Tel: 92-42-7324148(W) 6301954(H)
Fax: 92-42-7246393(W)

Specialisation : International Law.

Areas of Current Research:
International Law; Law of the Sea; Implementing Legislations; International Trade Laws.

Recent/Forthcoming Publications:
International Documents on Pakistan.

■ *Peshawar*

GHUFRAN, Nasreen (F/35)
Assistant Professor
Dept. of International Relations, Peshawar University
Peshawar, Pakistan.
Tel: 92-521-42363(W) 521-271940(H)
Fax: 92-521-273548(H)

Specialisation : International Relations.

Areas of Current Research:
Central Asia; Pakistan & Central Asia; Islam.

Main Work at Hand:
South Asian Refugees in the Post Cold War Era; Central Asian Refugees.

Recent/Forthcoming Publications:
"Pakistan, The Central Asian Republic & Economics", *The Journal of Humanities & Social Sciences,* Vol.2, No.1, March 1994, (Peshawar University); "The Role of Islam in Central Asia", *Strategic Perspectives,* Winter 94. Spring 95 No.1&2, (The Institute of Strategic Studies, Islamabad); "The Islam Factor in Pakistan's Relations with Central Asian Republics", *BIIS Journal,* Vol. 16, 1995; "Pakistan: Foreign Policy Toward Central Asia", *Area Study Centre,* (Bi-annual Research Journal, Peshawar University, 1995).

HILALI, Agha Z.A. (M/47)
Lecturer, Dept. of International Relations
University of Peshawar, Peshawar, Pakistan.

Specialisation : International Relations.

Main Work at Hand:
Security Imperatives of South Asia & it's Impact on Politics & Economy: A Case Study of Pakistan 1980-1988.

KHAN, Azmat Hayat (M/47)
Director, Area Study Centre (Central Asia)
Peshawar University, Peshawar, Pakistan.

Specialisation : Central Asia.

Areas of Current Research:
Afghanistan & the Five Central Asian Republics of the Former USSR.

Main Work at Hand:
Pakistan-Turkmenistan Commercial Links; Import Requirements of Uzbekistan.

Recent/Forthcoming Publications:
Islam in Central Asian Republics; Economic Re-integration of the Former USSR.; Water Problem in Central Asia.

QADEEM, Mussarat (F/35)
Dept. of Political Science, University of Peshawar
Peshawar, Pakistan.
Tel: 92-91-42292(W) 0521-842284(H)
Fax: 92-91-843534(H)

Specialisation : International Law.

Areas of Current Research:
South Asia; Women's Socio Political Emancipation; International Issues.

Recent/Forthcoming Publications:
Strategies for Conflict Resolution & Confidence Building Between India & Pakistan; "Iqbal's Message to the Muslims of 20th Century", *Journal of Rural Development,* (June 1996); Political Conciousness Among the Females in Pakistan; "Political Participation of Women in N.W.F.P.," *Women as Agents of Change* (Beijing, 1995); "The Role of Judiciary in the Constitutional Development of Pakistan" *Journal of Law & Society,* 1994; "Multinational Cooperations & State Control", *Journal of PARD*, April 1995.

RAUF, Abdul (M/33)
Research Associate, Pakistan Study Centre
University of Peshawar, Peshawar, Pakistan.
Tel: 92-91-42350(W)

Specialisation : Political Science.

Areas of Current Research:
History & Politics of the North West Frontier Province of Pakistan; Democracy & the Problem of Legitimacy in South Asia with Special Reference to Pakistan.

Main Work at Hand:
Khilafat Movement in N.W.F.P.; Muslim Politics in N.W.F.P. (1919-1930) with Special References to Pan-Islamic Ideas.

YUSUFZAI, Rahimullah (M/41)
Resident Editor, c/o The News
Peshawar Bureau, Saddar Road, Peshawar, Pakistan.

Specialisation : English, Political Science

Areas of Current Research:
Pakistan, Afghanistan, India, Nepal, Other South Asian Committees; Central
Asian Republics/Gulf States; Political & Strategic Issues in the Region;
Origin & Future of Pashtoon Tribals/Islamic Movements etc.

Recent/Forthcoming Publications:
Durrani-Ghalji Rivarly in Afghanistan; Bhutan - Struggle for an Independent
Foreign Policy; Strait of Hormuz - Lifeline of the Gulf.

■ *Rawalpindi*

BABAR, Aneela Zeb (F/24)
27-A Harley Street,, Lane No. 6
Rawalpindi, Pakistan.

Specialisation : Defence & Strategic Studies.

Areas of Current Research:
Confidence Building Measures (Military & Non Military); Role of
Inteligence Agencies in Security Policies; Nuclear Option in Security Policy.

Main Work at Hand:
Pakistan's Security, An Evaluation of Nuclear Option; Role of Mossad in
Israel Security Policy.

HUSAIN, Ross Masood (M/67)
88 Race Course Scheme, Race Course Road,
Street 3, Rawalpindi, Pakistan.
Tel: 92-51-563309(W) 852208(H)
Fax: 92-51-564244(W) 282151(H)
E-mail: syed%friends@sdnpk.undp.org

Specialisation : International Law.

Areas of Current Research:
South Asia; Central Asia; European Union.

Recent/Forthcoming Publications:
OIC & Contemporary Issues of the Muslim World, 1997; Pakistan Agenda
for Change 1997.

JUNAID, Shahwar (F)
327/c, Lane No.4, Peshawar Road,
Rawalpindi, Pakistan.
Tel: 92-51-472008(W)

Specialisation : Strategic Communications.

Areas of Current Research:
International & Current Affairs; The Global Economy & Defence Issues;
Communications; Media Policy.

Main Work at Hand:
Transnational & National Identity; National Defence Policy.

Recent/Forthcoming Publications:
Communications Media & Public Policy, (Sep 1995); Transnational &
National Identity.

MADNI, Mushtaq (M/63)
13, New Lalazar, Rawalpindi,
Pakistan.

Specialisation : Military Subjects.

Areas of Current Research:
Defence Subjects:; Social & Political Events, Social Welfare.

Main Work at Hand:
In Pursuit of Happiness Military Essays for Army Officers).

SADIQ, Foqia (F/24)
23/D, Lane 2, Tulsa Road
Lalazar, Rawalpindi, Pakistan.

Specialisation : Defence & Strategic Studies.

Areas of Current Research:
Peace & Security Issues.

Main Work at Hand:
Evolution of Peace Movements in the US.

SYED, Talat Hussain (M/29)
The News, Jang Building, Al-Rehman Plaza
Murree Road, Rawalpindi, Pakistan.
Tel: 92-51-555201(W)
Fax: 92-51-555371(W)

Specialisation : International Relations.

Areas of Current Research:
Nuclear Disarmament; Regional Trouble Spot; Regional Trade.

Main Work at Hand:
"Clinton's Kashmir Policy", (Chapter of an Edited Book); "Pakistan-US Relations Under Zia", (Part of an Edited Book).

PHILIPPINES

KUMAR, Rajiv (M/44)
Economist, Asian Development Bank,
P.O. Box 789, Manila 0980, Philippines.
E-mail: rkumar@mail.asiandevbank.org

Specialisation : Economics.

Areas of Current Research:
Trade and Balance of Payments; Individual Competitiveness; Technology.

HOSSAIN, Ishtiaq (M/43)
Dept. of Political Science, University of Singapore
10, Kent Ridge Crescent, Singapore 119260.
Tel: 65-772-6130(W)
Fax: 65-779-6815(W)
E-mail: polhossa@leonis.nus.sg

Specialisation : Political Science.

Areas of Current Research:
International Relations & Comparative Politics in South Asia & the Middle East.

Recent/Forthcoming Publications:
"Bangladesh: A Nation Adrift", *Asian Journal of Political Science,* Vol.3, No.1, Jun-1995; "An Experiment in Sustainable Human Development: The Grameen Bank of Bangladesh"; *Bangladesh & the Gulf War: Response of a Small State*, (Academic Press, Dhaka, Bangladesh).

SRI LANKA

■ Colombo

ABEYSEKARA, Irangani Manel (F/62)
184D Kannate Road, Thalapathpitiya
Nugegoda, Sri Lanka.

Areas of Current Research:
SAARC; Gender Issues.

ABEYWARDENE, Dickelle Lekamlage (M/54)
University of Jayawardenepura, Gangodawila
Nugegoda, Sri Lanka.
Tel: 94-1-826163(W)

Specialisation : International Relations.

Areas of Current Research:
International Relations in Sri Lanka; Contemporary History; Ancient Civilization.

Recent/Forthcoming Publications:
International Relations in Sri Lanka 1948-1983.

ARIYARATNE, R.A. (M/53)
Dean's Office, Faculty of Arts
University of Colombo, Colombo 03, Sri Lanka.

Specialisation : International Relations.

Areas of Current Research:
South Asian Politics; International Relations; Ethnic Studies.

Main Work at Hand:
Voice of America in Sri Lanka; Proposed Devolution of Power in Sri Lanka.

Recent/Forthcoming Publications:
Sri Lanka Peace Process in Theoratical Perspective; Reforms in the Teaching Methodologies.

ARIYASINHA, Ravinatha Pandukabhaya (M/34)
Director, Publicity & Spokesman, Ministry of Foreign Affairs
Colombo 01, Sri Lanka.

Specialisation : International Relations.

Areas of Current Research:
South Asian Politics; Sri Lanka Foreign Policy; Public Diplomacy/International Communication.

Main Work at Hand:
Economic Liberization in South Asia: It's Implications on Inter-State Relations.

Recent/Forthcoming Publications:
"Foreign Policy Decision Making in Sri Lanka"; "Sri Lanka & SAARC: Dilemma in Regionalism"; in *Aspects of Sri Lanka's Foreign Relations in a Changing World*, (BCIS); "Building an Enduring Indo-Sri Lanka Relationship", *in Society, Economy and Polity in Sri Lanka* (South Asia Studies Centre, Jaipur).

BASTIAMPILLAI, Bertram Emil Saint Jean (M/65)
Parliamentary Commissioner for Administration,
594/3, Galle Road, Colombo 03,
Sri Lanka.
Tel: 94-1-588982(W) 500581(H)

Specialisation : History.

Areas of Current Research:
Contemporary Politics; Foreign Affairs; Contemporary History of South
Asia.

Main Work at Hand:
Development & Environmental Protection in a Federal Set-Up; Devolution &
Power Sharing: Means to Peace & Development in Sri Lanka.

Recent/Forthcoming Publications:
Changing Fortunes of Indian Immigrants in Independent Sri Lanka; A
Decade of South Asian Regional Cooperation; The Ombudsman as an
Instrument of Good Governance.

BASTIAN, Sunil (M/47)
International Centre for Ethnic Studies
2, Kynsey Terrace, Colombo 08, Sri Lanka.
Tel: 94-1-698048(W)
Fax: 94-1-696618(W) 820079(H)

Specialisation : Geology.

Areas of Current Research:
Development Policy & Ethnicity; NGOs & Civil Society Politics.

Main Work at Hand:
Devolution & Development.

Recent/Forthcoming Publications:
The New Consensus on Development Assistance: Good Governance & Civil
Society Politics; Living Between Participation: & Self Determination:
Reflections on the Plantation Sector; "Control of State Land - The Devolu-
tion Debate" in *Devolution and Debate* (ICES 1996).

COORAY, M. George A. (M/51)
Co-ordinator, International Relations Programme
University of Colombo, Colombo 03, Sri Lanka.
Tel: 94-1-583810(W) 698836(H)

Specialisation : International Relations.

Areas of Current Research:
Arms Control & Disarmament; Conflict & Conflict Resolution; Security.

Main Work at Hand:
Security & Intervention; Militarization in South Asia.

Recent/Forthcoming Publications:
The Indo-Sri Lanka Peace Accord: A Critical Appraisal.

DE SILVA, Wimala (F/75)
Chairperson, National Committee on Women
11/1, Malalasekara Pedesa, Colombo 07, Sri Lanka.
Fax: 94-1-595841(W)

Specialisation : Education.

Areas of Current Research:
Women's Issues; Education.

Main Work at Hand:
A Study of Ageing in Sri Lanka - The Care of the Elderly with Special
Reference to Women; The Education of the Girl Child 1816-1995, A Case
History of Newstead.

Recent/Forthcoming Publications:
"The Refugee Child", in *Shadows & Vistas*; "Political Participation:
of Women in Sri Lanka, 1985-1995", *in Facts of Change*; The Impact
of the Competitive System of Education on Family Values.

DEWARAJA, Lorna Srimathie (F/66)
The Bandaranaike International Diplomatic Training Institute
Suite No.3G-07, BMICH, Bauddhaloka Mawatha, Colombo 07, Sri Lanka.
Tel: 94-1-682110(W)
Fax: 94-1-682111(W)

Specialisation : History.

Areas of Current Research:
South Asian History; Women's Studies; Muslim Minorities in Asia.

Main Work at Hand:
South Asian Women in Politics; The Indigenisation of the Catholic Church in
Sri Lanka.

Recent/Forthcoming Publications:

*The Kandyan Kingdom of Sri Lanka 1707-1782; Sri Lanka Through French
Eyes; The Muslims in Sri Lanka - One Thousand Years of Ethnic Harmony.*

EDRISINHA, Rohan Hemantha (M/37)
Centre for Policy Research & Analysis, Faculty of Law
University of Colombo, P.O.Box 1490, Colombo, Sri Lanka.
Tel: 94-1-595667(W)
Fax: 94-78-68297(W)
E-mail: cepra5@sri.lanka.net (W)

Specialisation : Law.

Areas of Current Research:
Constitutional Law; Human Rights; Comparative Politics.

Main Work at Hand:
Constitutionalism & the Judiciaries of Sri Lanka & South Africa:
A Comparative Survey; Comparative Federalism.

Recent/Forthcoming Publications:
The Freedom of Conscience of Members of Parliament &
Representative Democracy in Sri Lanka; Media Law of Sri Lanka.

ENDAGAMA, Malani (F/55)
Dept. of History & Archaeology, University of Sri Jayawardenepura
Gangodawila, Nugegoda, Sri Lanka.
Tel: 94-1-852695(W) 854255(H)

Specialisation : History.

Areas of Current Research:
Provincial & Local Government & Devolution; Women's Studies;
History of Constitutions - Comparative Studies.

Main Work at Hand:
Historical Background & Growth of the Constitution of USA.

Recent/Forthcoming Publications:
History of Gamsaba System in Sri Lanka.

FOSTER, Yolanda (F/25)
S.S.A., 425/15, Thimbirigasyaya Road
Colombo 05, Sri Lanka,

Specialisation : Development.

Areas of Current Research:
Social Integration or Exclusion; Politics of Identity; Globalisation.

Main Work at Hand:
Strategies of Exclusion: A Case Study of Dalit in India; Beyond The Gaze.

Recent/Forthcoming Publications:
Patriarchy on the Plantations.

GOONERATNE, John H.N. (M/59)
Associate Director
Regional Centre for Strategic Studies
4-101, BMICH, Bauddhaloka Mawatha
Colombo 7
Tel: 94-1-688601, Fax: 94-1-688602
E-mail: johng@sri.lanka.net

Specialisation: International Relations

Areas of Current Research:
Ethnic Conflict; Security Matters (Regional)

GUNASEKARA, Herath M. (M/68)
Friedrich Ebert Stiftung, 4, Adams Avenue
Colombo 04, Sri Lanka.
Fax: 94-1-862068(H)

Specialisation : Arts.

Areas of Current Research:
Journalism, Electronic Media; Mass Communication; Regional Media
Issues in SAARC Region.

GUNATILAKA, Ramani Sonali (F/29)
Institute of Policy Studies, 99, St. Micheal's Road
Colombo 03, Sri Lanka.
Tel: 94-1-431368(W)
E-mail: ips@srilanka.net (W)

Specialisation : Development Economics.

Areas of Current Research:
Labour Market Policies; Poverty Alleviation Issues; World Bank Conditionality.

Main Work at Hand:
Labour Legislation & the Impact on Labour Demand; Poverty
Alleviation Programmes Study.

Recent/Forthcoming Publications:
Long Term Trends in the Sri Lanka Economy.

GUNEWARDENA, Victor (M/68)
41, Pepiliyana Road, Nedimala
Dehiwala, Sri Lanka.
Tel: 94-1-714395(W)
Fax: 94-1-735910(W)

Areas of Current Research:
Press & Development in Sri Lanka; Violence in Sri Lanka.

Main Work at Hand:
Constraints on Press Freedom in Sri Lanka; Media Reach - Readers, Listners
& Viewers.

Recent/Forthcoming Publications:
Media an Aid to Unstructure Cooperation; Sri Lanka's Perspectives on Media
& Communication; "Press as Promoter" in *South Asian Regional Coopera-
tion*, FES, 1993.

JAYAWARDANE, Amal (M/48)
Dept. of History & Political Science, University of Colombo
Colombo 03, Sri Lanka.

Specialisation : Russian Studies.

Areas of Current Research:
International Politics in South Asia; Sri Lankan Foreign Policy; Human
Rights Issues.

Main Work at Hand:
The Soviet Attitude Towards the Indo-Sri Lanka Problems; Some
Constititional Aspects of Foreign Policy Making: USA & Sri Lanka; Finland
Vs. Sri Lanka: Use & Misuse of an Analogy.

JAYAWARDENA, Upali (M/63)
Marga Institute, 93/10, Dutugemunu Street
Kirillapone, Colombo 06, Sri Lanka.
Tel: 94-1-828544(W) 713259(H)
Fax: 94-1-828597(W)
E-mail: marga@sri.lanka.net

Specialisation : Public Administration.

Areas of Current Research:
Economic Studies; Social Studies; International Studies.

Main Work at Hand:
Brain Drain Problem; Community Water Supply & Sanitation.

Recent/Forthcoming Publications:
Economic Cooperation Between South Asia & Japan.

JAYAWEERA, Kapila Susantha (M/30)
Ministry of Foreign Affairs, Colombo 01
Sri Lanka,

Specialisation : International Relations.

Areas of Current Research:
South Asian Strategic Studies; South East Asia; Foreign Policy.

Main Work at Hand:
Foreign Policy Under J.R. Jayawardene; Indo-US Relations Since
1990-1995.

KANESALINGAM, Vaitilingam (M/79)
Friedrich-Ebert-Stiftung, Sri Lanka Office
4 Adams Avenue, Colombo 04.
Tel: 94-1-502710(W) 585909(H)

Specialisation : Public Finance.

Areas of Current Research:
Regional Coperation; International Trade & Development; Development Administration.

Main Work at Hand:
Economic Liberalisation.

Recent/Forthcoming Publications:
Private Sector & Regional Cooperation in South Asia; Collective Self-Reliance in South Asia; Economic Liberalisation in Sri Lanka 1995.

LAKSHMAN, W.D. (M/54)
University of Colombo, 94, Kumaratunga Munidasa Mawatha
Colombo 03, Sri Lanka.
Tel: 94-1-583810(W) 864943(H)
Fax: 94-1-583810(W)
E-mail: vc@admin.cmb.ac.lk

Specialisation : Economics.

Areas of Current Research:
Development Economics; International Economic Relations; Foreign Direct Investment; Industrial Policy & Stucture.

LEITAN, Tressie (F/62)
Dept. of History & Political Science, University of Colombo
Colombo 03, Sri Lanka.

Specialisation : Political Science.

Areas of Current Research:
Power Sharing/Devolution; Ethnicity & Conflict Resolution; Local Participation:.

Main Work at Hand:
Political Integration through Decentralization & Devolution of Power: The Sri Lankan Case; Centre-Province Relations Under the System of Provincial Council.

Recent/Forthcoming Publications:
Power to the Pheriphery: the Challenge of Power Sharing in Sri Lanka.

LOGANATHAN, Ketheshwaran (M/44)
Centre for Policy Research & Analysis
Faculty of Law, Reid Avenue, Colombo 03, Sri Lanka.
Tel: 94-1-595667(W)

Specialisation : Development Studies.

173

Areas of Current Research:
Conflict Resolution - Task of Nation Building & Constitutional
Reform; Sustainable Development; Role of State in Economic Development.

Main Work at Hand:
The Plantation System in Sri Lanka & Sustainable Development; Political
Context of Attempts at Mediation of the Ethnic Conflict in Sri Lanka.

Recent/Forthcoming Publications:
Experience of North East Provincial Councils with 13th Amendment
to the Constitution.

MENDIS, Vernon Lorraine Benjamin (M/70)
Director General
Bandaranaike International Diplomatic Training Institute
Suite 3G-07, BMICH, Colombo 07, Sri Lanka.
Tel: 94-1-682109(W) 501111(H)
Fax: 94-1-682111(W)

Specialisation : International Relations.

Areas of Current Research:
History, with Special Reference to Asia; Diplomatic History; Modern International Relations.

Main Work at Hand:
The Rulers of Sri Lanka; Foreign Relations of Sri Lanka post-1965.

Recent/Forthcoming Publications:
SAARC: Origins, Organisations Prospects (Centre for Indian Ocean Studies,
University of Western Australia, Perth, 1991); *National Security
Concepts of Sri Lanka* (UNIDI Publication, Geneva 1992.)

PERERA, Amrith Rohan (M/48)
Ministry of Foreign Affairs, Republic Square
Colombo 01, Sri Lanka.

Specialisation : Public International Law.

Areas of Current Research:
International Terrorism; International Criminal Jurisdiction; Law of the Sea.

Main Work at Hand:
Initiatives to Combat Terrorism & the Development and Analysis of
International Law.

Recent/Forthcoming Publications:
International Terrorism.

PERERA, Myrtle (F/60)
Marga Institute, 93/10
Dutugemunu Street, Colombo 06, Sri Lanka.

Specialisation : Economics.

Areas of Current Research:
Social Science, Gender Studies, Health, Environment.

Main Work at Hand:
Child Malnutrition; Pilot Project on Community, Seaweed Farming.

Recent/Forthcoming Publications:
Self Emploment Schemes for Women in Sri Lanka; Rural Poverty in
Sri Lanka: Priority Issues & Policy Measures; Impact of Macro
Policies on Social Structure in Women, Household & Change, Rural
Poverty in Developing Asia - Sri Lanka Country Side.

PERERA, Sasanka (M/33)
Department of Sociology, University of Colombo
Colombo 03, Sri Lanka.
Tel: 94-1-500452(W)
Fax: 94-1-583810(W)

Specialisation : Anthropology.

Areas of Current Research:
Political Violence; Ethnicity & Nationalism; Education in
Multicultural Society.

Main Work at Hand:
*Living with Torturers & Other Essays of Intervention: Sri Lankan Society,
Politics & Culture on Perspective* (Colombo: ICES, 1995).

RAJAPAKSE, Puran (M/39)
Institute of Policy Studies, 99, St. Micheal's Road
Colombo 03, Sri Lanka.
Tel: 94-1-574983(H)
Fax: 94-1-573083(H)

Specialisation : Economics.

Areas of Current Research:
Monetary Economics; Trade Policy; Capital Market.

Main Work at Hand:
Impact of the Uruguay Round on Sri Lanka; Can the Government Control
Money Supply: An Empirical Analysis of the Determinants of Money
Growth in Sri Lanka, 1970-1989.

Recent/Forthcoming Publications:
Would a Reduction in Trade Barriers Promote Intra - SAARC Trade.

SAMARASINGHE, Dilip Susruta (M/35)
Air Lanka Ltd., 37, York Street
Colombo 01, Sri Lanka.
Tel: 94-1-731711(W) 696068(H)
Fax: 94-1-735122(W)

Specialisation : International Affairs.

Areas of Current Research:
International Conflicts in the Cold War & Post Cold War Eras; Security;
Weapons (especially Aviation & Missile) & Strategy.

Main Work at Hand:
Jet Fighters: South Asia's Latest Arms Race; Sino-Pakistani Cooperation in
the Development of Military Aircraft.

Recent/Forthcoming Publications:
Airline Franchising; The Development of New Aircraft Types & Their
Impact on the Airline Industry; International Security at the End of the Cold
War.

SANDERATNE, Nimal E.H. (M/58)
Chairman, National Development Bank
40, Navam Mawatha, Colombo 02, Sri Lanka.
Tel: 94-1-447474(W) 501406(H)
Fax: 94-1-440262(W)

Specialisation : Development Studies.

Areas of Current Research:
Macro Economics; Development Economics; Environmental Economics.

Main Work at Hand:
Putting People First: Sri Lanka's Development Experience, Inflation &
Economic Growth in Sri Lanka.

Recent/Forthcoming Publications:
*Political Commitment & Economic Sustainability: Early Achievements &
Later Strains in Social Development, Sri Lanka; Sri Lanka: Democracy &
Accountability in Decline.*

SARAVANAMUTTU, Paikiasothy (M/37)
Executive Director
Centre for Policy Alternatives
32/3, Sir Earnest de Silva Mawatha (Flower Road)
Colombo 07, Sri Lanka.

Tel: 94-1-565305/565304(W) 597561(H)
Fax: 94-1-074-714460(W)
E-mail: cpa@sri.lanka.net

Specialisation : International Relations.

Areas of Current Research:
Security in the Developing World; International Politics in South Asia;
International Political Theory.

Main Work at Hand:
Oral History of Ethnic Conflict in Sri Lanka; Conflict Resolution with
Reference to Ethnic Conflict & Use of Costitutional Reform as a Conflict
Resolution Mechanism.

Recent/Forthcoming Publications:
Problems & Prospects for a Regional Security System in Asia (BUS 96);
"Minorities in South Asia", *International Consultation on Minority Protec-
tion*, (ICES 93); *Safeguarding Minority Communities in South Asia.*

SATHANANTHAN, Sachithanandam (M/51)
MANDRU (Institute for Alternative Development & Regional Cooperatiion)
62/7, Dabare Mawatha, Colombo 05, Sri Lanka.
Fax: 94-1-590473(W)

Specialisation : Land Economy.

Areas of Current Research:
Nationalism; Democratization & the State in South Asia.

Main Work at Hand:
"On South Asia", in *Choosing Our Future: Visions of a Sustainable World*
(World Resources Institute, 1995); The Elusive Dove: An Assesment of
Conflict Resolution Initiatives in Sri Lanka.

Recent/Forthcoming Publications:
Anti Federalism, National Questions & Confederation: Political Crisis in Sri
Lanka.

SATHISCHANDRA, D.H. (M/59)
Marga Institute, 93/10
Dutugemunu Street, Colombo 06, Sri Lanka.
Tel: 94-1-828545(W) 811046(H)
Fax: 94-1-828597(W)

Specialisation : Economics.

Areas of Current Research:
Trade; Environment; Labour Relations, Labour Marketing & Institutions.

Main Work at Hand:
Health Accountability in the Industrial Sector; South Asia Japan Cooperation.

Recent/Forthcoming Publications:
Industrial Cooperation in South Asia; Opportunities & Strategies in Selected Products.

SELVAKKUMARAN, Naganathan (M/42)
Senior Lecturer in Law, Faculty of Law
University of Colombo, Colombo, Sri Lanka.
Tel: 94-1-591945(H)

Specialisation : Law.

Areas of Current Research:
Administrative Law; Constitutional Law; Local Government Law.

Main Work at Hand:
The Role of the Wist of Mandamns in the Administrative Law of Sri Lanka.

Recent/Forthcoming Publications:
Mass Media Laws & Regulations in Sri Lanka; Essays on Constitutional Law.

SENADHIRA, Sugeeswara Premalal (M/48)
56, Siripura, Hokandara Road
Talawatugoda,
Sri Lanka.
Tel: 94-1-868196, Fax: 94-1-868196

Specialisation: Media Studies

Areas of Current Research:
Indo-Lanka Relations; Press Freedom; Ethnic Conflicts

Main Work at hand:
A Comparative Study of the Tamil Rebel Movement and Sikh Militancy;
Freedom of Expression in Post-Premadasa Sri Lanka.

Recent/Forthcoming Publication:
President Jayawardene's National and International Policies (New Delhi, 1985); *Under Siege - Mass Media in Sri Lanka* (New Delhi, 1996).

TILAKARATNE, Dhananjaya (M/32)
5, Jayaratne Avenue, Colombo 05
Sri Lanka.

Specialisation : International Relations.

Areas of Current Research:
Human Rights; International Human Rights Law.

WEERAKOON, Dushni (F/29)
Institute of Policy Studies, 99, St. Michael's Road
Colombo 03, Sri Lanka.

Specialisation : Economics.

Areas of Current Research:
Trade & Investment in South Asia; International Economics;
Macro Economics.

Main Work at Hand:
Scheme of Regional Cooperation in South Asia: The Role of Japan.

Recent/Forthcoming Publications:
How Open Has the Sri Lanka Economy Become? Trends in Trade & Trade
Taxes; Japanese Role in Asian Development: Lessons for Sri Lanka.

WICKRAMASINGHE, Nira (F/31)
Dept. of History & Political Science
Reid Avenue
Colombo 03, Sri Lanka.
Tel: 94-1-500433(W)

Specialisation : Philosophy, History.

Areas of Current Research:
Nationalism; Ethnicity; Human Rights.

Main Work at Hand:
Ethnic Politics in Colonial Sri Lanka; Introduction to Social Theory.

Recent/Forthcoming Publications:
From Human Rights to Good Governance - The Aid Regime in the 1990s,
(Berg Press, Oxford, Spring 1996); "The Return of Keppetipola's Cranium -
Authority in a Modernising Nation", *Economic and Political Weekly*, 1996.

■ *Kandy*

DE SILVA, Kingsly M. (M/64)
Executive Director
International Centre for Ethnic Studies
554/1 Peradeniya Road, Kandy, Sri Lanka.
Fax: 94-8-234892(W)
E-mail: ices@slt.lk

Specialisation : History of Sri Lanka.

Areas of Current Research:
Ethnic Conflicts; Language Policies.

Main Work at Hand:
J.R. Jayawardena of Sri Lanka: A Political Biography (With Howard Wriggins); University of Sri Lanka, History of Sri Lanka, Vol.2.

Recent/Forthcoming Publications:
Regional Powers & Small State Security: India in Sri Lanka, 1977-1990, (Woodrow Wilson Centre Press & Johns Hopkins, University Press, Washington D.C., 1995); *Transfer of Power in Sri Lanka* (British Documents of the End of Empire Series, Her Majesty's Stationary Office, London).

■ *Peradeniya*

DE ZOYSA, M.O.A. (M/50)
Dept. of Political Science, University of Peradeniya
Peradeniya, Sri Lanka.

Specialisation : International Relations.

Areas of Current Research:
Regional Cooperation - SAARC & ASEAN; Constitutional Studies.

Main Work at Hand:
Civil Religion Thesis & the Constitution Making in Sri Lanka.

Recent/Forthcoming Publications:
"Politicization", *Sociology Magazine*, (University Press); "Centre-State Relations in India", *Political Science Magazine*, (University Press).

LIYANAGE, Kamala (F/44)
Dept. of Political Science, University of Peradeniya
Peradeniya, Sri Lanka.
Tel: 94-8-88301(W) 88301(H)

Specialisation : Political Science.

Areas of Current Research:
Ethnic & Racial Studies; Comparative Government & International Relations; Women's Studies.

Main Work at Hand:
"Friends & Foes of the Indo-Sri Lanka Accord"; "Women in Higher Education", in *Perspectives from University of Peradeniya* (CENWOR, 1995).

Recent/Forthcoming Publications:
Party Women: Their Role in Sri Lankan Politics (ICES, Kandy, 96); *Human Rights in Sri Lanka; Feminist Movement in Sri Lanka*, (Stanford University, USA, 95).

SIVARAJAH, Amabalavanar (M)
Dept. of Political Science, University of Peradeniya
Peradeniya, Sri Lanka.

Specialisation : Ethnic Conflict.

Areas of Current Research:
Ethnic Conflict & Conflict Resolutions; South Asia in International Politics;
Foreign Policy of Sri Lanka.

Main Work at Hand:
Minorities & Constitution Making in Sri Lanka; Role of the Theory of
Traditional Home Lands of Tamils in Sri Lanka in the Tamil Nationalism in
Sri Lanka.

Recent/Forthcoming Publications:
US-Sri Lanka Relations in the 1980s (Kandy:1995); *Politics of Tamil Nationalism in Sri Lanka; Ethnic Conflict in Sri Lanka: A Model for Peaceful
Resolution.*

UKRAINE

MOSTAFA, Golam (M/41)
Dept. of International Affairs, University of Kiev Mohyla Academy,
2, Skovoroda Street, Kiev 254070, Ukraine.
Fax: 380-613-226-2027(H)

* **Specialisation** : Political Science.

Areas of Current Research:
Political Theories, Comparative Political Systems.

Main Work at Hand:
National Interest & Foreign Policy, (South Asian Publishers, 1995); Gulf
War & the New World War, (BIISS, 1993).

Recent/Forthcoming Publications:
Post-Soviet Central Asia: Changes & Directions.

ALI, Syed Mahmud (M/43)
BBC World Service, Bush House
The Strand, London WC2B 4PH, United Kingdom.
Tel: 44-171-257-2300(W)
Fax: 44-171-379-0975(W)

Specialisation : War Studies.

Areas of Current Research:
State-Formation, National Integration, Ethno-National Centrifugal Tenden-
cies in Post-Colonial South Asia; Evolution of Political & Economic Struc-
tures & Roles of the State in the Late 1900's; Regional & International
Security Issues, Especially Nuclear Proliferation.

Main Work at Hand:
Strategic Coercion in South Asia; Elite Insecurity Perception in Bangladesh
Vis-a-Vis India.

Recent/Forthcoming Publications:
The Fearful State: Power, People & Internal War in South Asia, (London,
1993); *Civil Military Relations in the Soft State: The Case of Bangladesh*,
(Bath, University of Bath, 1994); A Chapter in Habib Zafarullah (ed.), *The
Zia Episode in Bangladesh*, (Delhi, 1996).

ELLNER, Andrea (F/33)
The Graduate School of European & International Studies
University of Reading, Whiteknights,
P.O.Box 218, Reading RG6AA
United Kingdom.
Fax: 44-173-4755442(W), 181-8028548(H)

Specialisation : Political Science.

Areas of Current Research:
Regional Security of Central & South Asia; Non Proliferation of Weapons of
Mass Destruction; Maritime Defence.

Main Work at Hand:
Whither Transition - Development & Security in Central Asia.

Recent/Forthcoming Publications:
Fissile Material Safety & Security in the Former Soviet Union (Adelphi
Paper, Oct. 1996); Nascent Security Regimes for Central & South Asia -
The Nuclear Dimension.

LATTER, Richard (M/47)
Deputy Director, Wilton Park
Wiston House, Steyning, West Sussex, UK.

Specialisation : US Policy in Middle East.

Areas of Current Research:
Security Issues; Arms Control; Asia.

Main Work at Hand:
South East Asian Security; Preventing the Proliferation of Weapons of Mass Destruction.

Recent/Forthcoming Publications:
"Ballistic Missile Proliferation in the Developing World", *Jane's Defence Magazine*, 1996; "Preventing the Proliferation of Biological Weapons", *Wilton Park Paper 109*.

MALIK, Iftikhar H. (M/46)
School of Historical Studies, Bath Spa University College
Bath, UK
Tel: 44-1225-875592(W)

Specialisation : International History.

Areas of Current Research:
Post-1947 Indo-Pakistan Relations - Historical Strategic;
US-South Asian Relations; Islam & the West.

Main Work at Hand:
Islam, Nationalism & the West; State & Civil Society in Pakistan; US-South Asian Relations.

Recent/Forthcoming Publications:
Islam, Nationalism & the West; *Islam, Nationalism & Nation Building in Pakistan* (USA); *Indo-Pakistan Relations: Hope or Despair*.

SAMAD, Yunas (M/44)
Dept. of Social & Economic Studies, University of Bradford
Richmond Road, Bradford, West Yorkshire, BD7 1DP, UK.

Specialisation : History.

Areas of Current Research:
Identity Politics in South Asia; South Asians in Europe; Globalization, Media & Identity.

Main Work at Hand:
Multiculturalism, Muslims & the Media; Muhajir Identity, Politics & the State.

Recent/Forthcoming Publications:
Nation in Turmoil: Nationalism & Ethnicity in Pakistan 1937-1958, (New Delhi: Sage 1995); *Culture, Identity & Politics: Ethnic Minorities in Britain*,

(Co-ed. T. Ranger & O. Stuart, Aldershot: 1996); "Pakistan or Punjabistan: Crisis of National Identity", *International Journal of Punjab Studies,* New Delhi: Sage.

SCHOFIELD, Victoria (F/45)
48, Westbourne, Park Road
London W25PH, United Kingdom.

Specialisation : History.

Areas of Current Research:
Sub-Continent.

Main Work at Hand:
Bhutto: Trial & Execution; Every Rock, Every Hill: North-West Frontier & Afghanistan.

Recent/Forthcoming Publications:
Kashmir in the Crossfire, (I.B. Tauris, 1996).

SMITH, Chris (M/40)
Centre for Defence Studies, King's College
Strand, London WC2R 2LS, United Kingdom.

Specialisation : International Relations.

Areas of Current Research:
Nuclear Proliferation in South Asia; Light Weapon Proliferation in South Asia; Conventional Defence in South Asia.

Main Work at Hand:
The Proliferation of Light Weapons in South Asia.

Recent/Forthcoming Publications:
"The Impact of Light Weapons on Security: A Case Study of South Asia", *Sipri Year Book, 1995*.

TAYLOR, David (M/50)
School of Oriental & African Studies, Russell Square
London WC1H 0XG, United Kingdom.
Tel: 44-171-323-6123(W)
Fax: 44-171-323-6020(W)
E-mail: dtl@soas.ac.uk

Specialisation : Political Science.

Areas of Current Research:
Indo-Pakistan Relations; Hindu Nationalism & the BJP.

O'NEIL, Robert
   Chichele Professor of the History of War
   All Souls College
   Oxford OX1 4AL,UK
   Tel: 44 -1865- 279385(W); Fax: 279299 (W)
   E-mail: robert.oneill@all-souls.oxford.ac.uk

   Specialisation: International Relations

   Areas of Current Research:
   New Security Threats; Nuclear Weapons

   Main Work at Hand:
   Papers relating to recent conflicts

UNITED STATES OF AMERICA

ABRAHAM, Itty (M/35)
   810 Seventh Avenue, New York, NY 10019
   USA.

   Specialisation : Political Science.

   Areas of Current Research:
   Political Economy of Arms Production; Culture, Norms & International
   Relations; State & Development.

   Main Work at Hand:
   The Colaba Programme: Atoms, Science & the State in India; Modernity &
   Nationalism in the Post Colonial State.

   Recent/Forthcoming Publications:
   "Science & Power in the Post Colonial State", *Alternatives,* Summer 1996;
   "Towards a Reflexive Security Studies, Weinbaum & Kumar", (eds.) *Asia
   Approaches the Millenium,* 1995.

BOUTON, Marshall M. (M/53)
725, Park Avenue, New York, NY 10021
USA.

Specialisation : Political Science.

Areas of Current Research:
South Asian Politics; South Asian International Relations; US-South Asian
Relations.

Recent/Forthcoming Publications:
Agrarian Radicalism in South Asia, India Briefing, 1987, 1995 (co Editor).

COHEN, Stephen Philip (M/59)
University of Illinois, 359, Armory Building
Champaign, IL 61820, USA.
Tel: 1-217-333-7086(W)
E-mail: s-cohen1@uiuc.edu

Specialisation : Political Science.

Areas of Current Research:
South Asian Security & Arms Control Issues; US Foreign Policy; Prolifera-
tion of Weapons of Mass Destruction; Role of the Military in Society and
Politics.
Recent/Forthcoming Publications:
The Indian Army (rev. edn. 1990); *The Pakistan Army*, 1985; Co-author,
Beyond Brasstacks; Editor of several books on South Asian security issues.

CORTRIGHT, David (M/49)
Joan B.Kroc Institute for International Peace Studies
University of Notre Dame, Notre Dame IN 46556, USA.
Tel: 719-631-8536(W)
E-mail: dcortright@igc.opc.org

Specialisation : Political Science.

Areas of Current Research:
Nuclear Weapons Policy/Disarmament; Economic Sanctions & Incentives.

Recent/Forthcoming Publications:
India & the Bomb: Public Opinion & Nuclear Options, Notre Dame Press,
1996; *Carrots & Cooperation in South Asia*; *Pakistan & the Bomb*.

DIEHL, Paul F. (M/37)
Dept. of Political Science, 361, Lincoln Hall
702 S. Wright Street, University of Illinois
Urbana IL 61801, USA.

Tel: 1-217-333-3881(W)
Fax: 1-217-244-5712(W)
E-mail: pdiel@uiuc.edu

Specialisation : Political Science.

Areas of Current Research:
Causes of War; UN Peacekeeping.

Main Work at Hand:
International Peacekeeping; Territorial Changes & International Conflict.

Recent/Forthcoming Publications:
The Dynamics of International Rivalries; The Politics of Global Governance.

GANGULY, Sumit (M/40)
Dept. of Political Science, Hunter College /CUNY
695 Park Avenue, New York NY 10021, USA.
Tel: 1-212-772-5666(W)
Fax: 1-212-650-3669(W) 212-866-4499(H)

Specialisation : Political Science.

Areas of Current Research:
Defence Policies in South Asia; Ethnic Violence in South & South East Asia;
Nuclear Proliferation.

Recent/Forthcoming Publications:
The Origins of War in South Asia (Boulder: Westview, 1994); *India Votes*
(Boulder: Westview, 1993); *Between War & Peace: The Kashmir Question
Revisited* (The Woodrow Wilson Centre Press); *Managing Ethnic Relations
in South & South East Asia*, with Micheal Brown ed., (M.I.T. Press).

HANSON, Elizabeth Crump (F/62)
Political Science Dept., Box U-24, 341 Mansfield Road
University of Connecticut, Storrs, CT 06269-1024, USA.
Tel: 1-860-486-2534(W)
Fax: 1-860-486-3347(W)
E-mail: hanson@uconnvm.uconn.edu (W)

Specialisation : Political Science; International Relations.

Areas of Current Research:
Public Opinion & Foreign Policy: India/US; South Asian Regional Relations;
Media/Foreign Policy.

Main Work at Hand:
The Structure of Foreign Policy Attitudes in India; Public Opinion & Foreign
Policy in India.

Recent/Forthcoming Publications:
"The Times of India in Changing Times", *Political Communication*, Oct/Dec 1995; "International News After the Cold War", *Political Communication*, Oct/Dec 1995; "The Global Media System & International Relations", in Kanti Bajpai & Harish C. Shukal (eds.), *Interpreting World Politics*, (Sage 1995).

HUSSAIN, Rifaat (M/42)
2315, Massachusetts Avenue, N.W.
Washington DC 20008
USA.

Specialisation : International Studies.

Areas of Current Research:
International Security; South Asia; Arms Control.

Main Work at Hand:
Rethinking Security: Case of South Asia; Pakistan's Role in the Changing International System.

Indo-Sri Lankan Relations; Indian Politics & Society; International Relations.

Main Work at Hand:
Post-colonial Insecurities: India, Sri Lanka & the Question of Eelam, (Book Manuscript).

Recent/Forthcoming Publications:
"Cartographic Anxiety: Mapping the Body Politic in India", Hayward Alker & Micheal Shapiro eds., *Territorial Identities & Global Flows*, (Minnesota, 1995); "The Importance of Being Ironic", *Alternatives* 18, 1993; "Political Culture, Constitutionalism & Democracy in India".

KHAN, Zillur Rahman (M/57)
Chairperson, Dept. of Political Science
University of Wisconsin, Oshtosh WI 54901, USA.
Tel: 1-414-424-0924(W)

Specialisation : Political Science.

Areas of Current Research:
Public Policy Making Process & the Complexity of Implementation; Sovereignty, National Interest & Regional Cooperation; Constitutional Growth & the Democratizing Process..

Recent/Forthcoming Publications:
Leadership Crisis in Bangladesh (University Press, 1984); *SAARC & the Superpowers* (University Press, 1991); *Leadership in the Developed Nation*

(Syracuse University, 1983); *Constitution & Constituional Issues* (University Press, 1986); *The Third World Charismat.*

KREPON, Micheal (M)
Henry L. Stimson Centre, 21 Dupont Circle, NW
5th Floor, Washington DC 20036, USA.
Tel: 1-202-223-5969(W)
Fax: 1-202-785-9034(W)
E-mail: krepon@stimson.org

Areas of Current Research:
South Asian Regional Security & Confidence Building; Nuclear Non Proliferation; International Security Issues.

Main Work at Hand:
Crisis Prevention, Confidence Building & Reconciliation in South Asia, (Vaguard Publishers, Pakistan, Manohar Publishers, India), *A Handbook of Confidence-Building Measures for Regional Security.*

KRISHNA, Sankaran (M/35)
Dept. of Political Science, University of Hawaii at Manoa
Honolulu, Hawaii 968221, USA.

Specialisation : Political Science.

Areas of Current Research:
Indo-Sri Lankan Relations; Indian Politics & Society; International Relations.

Main Work at Hand:
Post-colonial Insecurities: India, Sri Lanka & the Question of Eelam, (Book Manuscript, Forthcoming 1997).

Recent/Forthcoming Publications:
"Cartographic Anxiety: Mapping the Body Politic in India", Hayward Alker & Micheal Shapiro eds., Territorial Identities & Global Flows, (Minnesota, 1995); "The Importance of Being Ironic", *Alternatives 18, 1993*, Political Culture Constitutionalism & Democracy in India, Baurs & Franklin (eds).

KUX, Dennis (M/64)
125, Duke Street
Alexandria, Virginia 22314, USA.

Specialisation : International Affairs.

Areas of Current Research:
United States and South Asia.

Recent/Forthcoming Publications:
History of USA-Pakistan Relations.

LAMBA, Harinder Singh (M/47)
6921, Creekside Road, Downers Grove
Illinois 60516-3434, USA.
Tel: 1-708-964-2258(W)

Specialisation : Engineering Mechanics.

Areas of Current Research:
Ecological Rejuvenation; Participatory Democracy; Harmony Among Nation
& People.

Main Work at Hand:
Rural Development in India-Theory & Practice, 1985; "Call to Action" at the
Earth Summit, 1992.

LYNN, John Albert (M/52)
Dept. of History, University of Illinois at Urbana Champaign
309 Gregory Hall, 810 South Right Street, Urbana, Illinois 61801,USA.

Specialisation : History.

Areas of Current Research:
French Military History, 1610-1815; Eras of Military Change;
War in South Asia.

Recent/Forthcoming Publications:
Logistics in Western Warfare from the Middle Ages to the Present,
(Boulder, Colorado: Westview Press, 1993); The Bayonets of the Republic:
Motivation & Tactics in the Army of Revolutionary France, 1791-94,
(Urbana, Illinois: University of Illinois Press, 1984); *Giant of the Grand
Cycle: The French Army, 1610-1715* (Cambridge University Press, 1997).

MBAKU, John Mukum (M/45)
Dept. of Economics, Weber State University
3807 University Circle, Ogden, UT 84408-3807, USA.
Tel: 1-801 626 7442(W)
Fax: 1-801 626 6191(W)
E-mail: jmbaku@weber.edu (W)

Specialisation : Economics.

Areas of Current Research:
Public Choice; Development; Africa & Asia.

Main Work at Hand:
Preparing the Third World for the Twenty-First Century; Corruption &
Institutional Reforms in Developing Socities.

Recent/Forthcoming Publications:
"Preparing Africa for the Twenty-First Century: Lessons from Constitutional
Economics", *Constitutional Political Economy*, (1995); "Female Headship,
Feminization of Poverty & Welfare" (With M.S. Kimenyi), *Southern Eco-
nomic Journal*, 62 (1), 1995; "Military Intervention in African Politics:
Lessons from Public Choice", *Konjunkturpolitik* (Germany) 41 (3), 1995.

RAJAGOPALAN, Swarna (F/31)
Dept. of Political Science, 36-1 Lincoln Hall, 702 S. Wright Street
Urbana IL 61801, USA.
Tel: 1-217-333-3881(W)
Fax: 1-217-204-5157(W)
E-mail: srajagop@students.uiuc.edu

Specialisation : Political Science.

Areas of Current Research:
State-Building Issues & South Asian Security; Gender & International
Security/Relations; Traditional Indian Political Thought.

Main Work at Hand:
Conflicting Imperatives: Integration Versus Identity.

RIZVI, Hassan Askari (M/50)
Southern Asian Institute, Columbia University
International Affairs Building, 420 West 118th Street, #1133,
NY 10027, USA.
Tel: 1-212-854-8825(W)
Fax: 1-212-854-6987(W)

Areas of Current Research:
Foreign & Security Policies; Domestic Politics of Pakistan; Central Asia.

Recent/Forthcoming Publications:
The Military & Politics in Pakistan, (Lahore 1986); *Pakistan & the
Geostrategic Environment*, (London Macmillan 1993).

SARDESAI, Shonali (F/26)
Dept. of Political Science
University of Illinois, 505, South Busey, 404
Urbana, Illinois 61801.
Tel: 217-3840753(W)

Specialisation : International Relations.

Areas of Current Research:
Ethnicity; Conflict Resolution; Religion.

Recent/Forthcoming Publications:
ACDIS Occassional Paper.

WEINBAUM, Marvin G. (M/60)
361 Lincoln Hall, 702 So., Wright Street
Urbana, Illinois 61801, USA.

Specialisation : Political Science.

Areas of Current Research:
Liberalism Politics & Economics in the Islamic World; South Asian Regional Politics; Pakistan, Domestic Politics & Foreign Relations.

Recent/Forthcoming Publications:
Pakistan & Afghanistan: Resistance & Reconstruction, 1994; *South Asia Approaches the Millennium,* 1995, Co-edited with Chetan Kumar.

Part II
INSTITUTIONS

● Indian Ocean Centre (IOC)

Curtin University
GPO Box U1987
Perth 6001
Australia.

Chief Executive: Kenneth Mcpherson

Number of Professional Staff: 2M, 1F

Areas of Research: Indian Ocean Region: Security, Economics, Education

Main Work at Hand: Vocational and Higher Education in the Region; A Regional Security Dialogue; Second Track Processes of Regional Cooperation

Publications: - Journal: Indian Ocean Review (Quarterly); Policy Papers; Network News; Occasional Papers; Books.

Library Collections: - Periodicals: 30
- Newspapers: 14

Year of Establishment: 1995

Status: University Research Centre

● Peace Research Centre (PRC)

Research School of Political and Asian Studies
Australian National University
Canberra ACT 0200
Australia.

Chief Executive: Ramesh Takur

Number of Professional Staff: 5M

Areas of Research: Nuclear and Missile Proliferation; United Nations Peacekeeping in the 1990s; Ethnic Conflict

Main Work at Hand: Nuclear-Weapon-Free Zones; United Nations Peace Keeping in the 1990s; Ethnic Conflict

Publications: - Journal: Pacific Research, 4/year; Occasional Papers; Books.

Seminars and Conferences: - Frequency: Monthly
- Participation: 20

-Title of a Recent Event:
"World Court Project on Nuclear Weapons."

Library Collections:	- Books: 700
	- Periodicals: 150
Year of Establishment:	1984
Status:	Governmental Organisation.

BANGLADESH

■ *Chittagong*

● **Department of Economics**
University of Chittagong, Chittagong
Bangladesh.
Tel: 880-31-210131-44

Chief Executive:	Begum Syeda Tahera, Chairperson
Number of Professional Staff:	19M, 1F
Areas of Research:	Development Economics; Rural Economics.
Main Work at Hand:	Rural Economics Development Projects by Individual Teachers.
Publications:	- Journal: *Arthoniti* (Economic Journal)
	- Recent/Forthcoming Titles: *Bishaw Banijjo Lendener Arthoniti* (Economics of World Trade & Payments)
Seminars & Conferences:	- Frequency: Bi-monthly
	- Participation: 50
Library Collections:	- Books: 500
	- Periodicals: 10
Year of Establishment:	1968
Status:	Autonomous

■ *Dhaka*

● **Bangladesh Centre for Advanced Studies (BCAS)**
No: 620, Road No:10A, Dhanmondi,
Dhaka, Bangladesh
Tel: 880-2-815829 Fax: 811344

Chief Executive:	Saleemul Haq, Executive Director
Other Key Personnel:	A. Atiq Rahman, Director
Number of Professional Staff:	45M, 20F
Areas of Research:	Environment; Climate Change; Economic Development
Main Work at Hand:	National Environmental Management Action Plan; Climate Change Country Strategy; Regional Environment Study.
Publications:	Bangladesh Environment Newsletter (Quarterly); Books.

Seminars and Conferences:		
	- Frequency	Weekly
	- Participation	20-100
	- Title of a Recent Event: "NEMAP Workshop."	

Library Collections:		
	- Books:	10000
	- Periodicals:	200
	- Newspapers:	40

Year of Establishment:	1984
Status:	Non-Governmental Organisation

● **Bangladesh Environmental Lawyers Association (BELA)**
No. 47, Road No; 5, Dhanmodi Road
Dhaka 1205, Bangladesh
Tel: 880-2-864283, 868706 Fax: 862957

Chief Executive:	Secretary General
Other Key Personnel:	K.A.A. Quamruddin, Chairman
Number of Professional Staff:	7M, 9F
Areas of Research:	Environmental Law; Policy Analysis; Institutional Issues
Main Work at Hand:	Basic Research, Public Interest Litigation, Awareness Campaign
Publications:	Occasional Papers; Books; Newsletters

Seminars and Conferences: - Frequency   Monthly
- Title of a Recent Event:
"Professional Workshop of Lawyers on National Environmental Management Action Plan."

Scholarship,Fellowship,Training etc.:

    - Title of Programme: Environmental Law
    - Number: 30
    - Duration: 1 week
    - Eligible: No Restriction
    - Funded by: The Ford Foundation

Library Collections: - Books: 1708
    - Periodicals: 27
    - Newspapers: 6

Year of Establishment: 1991

Status: Non-Governmental, Non-Profit, Non-Partisan Organisation

● **Bangladesh Institute of International and Strategic Studies (BIISS)**
1/46, Elephant Road,
Dhaka-1000,
Bangladesh.
Tel: 880-2-406234, 9347915 Fax: 832625
E-mail: biiss@bd.drik.net

Chief Executive: Director General

Other Key Personnel: Chairman
Abdur Rob Khan, Research Director

Number of Professional Staff: 14M, 6F

Areas of Research: Foreign Policy, Security and Strategic Issues, with Specific Relevance to Bangladesh; Politics Development and Inter-State Relations in South Asia and Regional and International Cooperation; Conflict Studies, Security of Small States, NPT and Environment, Resources and Sustainable Development.

Main Work at Hand: SAARC-Japan Study on Trade and Investment Cooperation in South Asia; National Capacity Building on the Bay of Bengal Coastal Zone Management, BIISS-OIC Collaborative Research Project; Dialogue Series on Sub-Regional Cooperation of Bangladesh with its Neighbouring States of India.

Publications:	- Journal:	BIISS Journal, (Quarterly); Bangladesh Foreign Policy Survey, (Monthly); Southeast Asia Monitor, (Forthcoming Quarterly).

-Recent/Forthcoming Titles:
Bangladesh and SAARC; Issues, Perspectives and Outlook; Bangladesh: Society, Polity, Economy; South Asia s Security: Primacy of Internal Dimension.

Seminars and Conferences:	- Frequency:	Varies
	- Participation:	150-200

-Title of a Recent Event:
International Seminar on Sharing of Ganges water Resources in South Asia: A Case of Bangladesh.

Scholarship, Fellowship, Training etc:

- Tile of Programme:	
BSIR Fellowship (M.Phil and Ph.D. Programme);	
- Subjects	Multi-disciplinary
- Number:	6
- Duration:	M.Phil-2 years/Ph.D-3 years
- Eligible:	Nationals of Bangladesh
- Funded by:	Ford Foundation

Library Collection:	- Books:	5000
	- Periodicals:	150
	- Magazines:	50

Year of Establishment:	1978
Status:	Autonomous Organisation

● Bangladesh Institute of Law and International Affairs (BILIA)

No: 22, Road No:7, Dhanmondi R/A, Dhaka-1205
Bangladesh
Tel: 880-2-9111718 Fax: 326303

Chief Executive:	Fakhruddin Ahmed, Honorary Director
Other Key Personnel:	Syed Ishtiaq Ahmed, Chairman

Number of Professional Staff: 3M

Areas of Research:	Studies in Law and International Affairs; Research in Law and International Politics.
Main Work at Hand:	A Framework and Implementation of Human Rights in Bangladesh.

Publications:

- Journal:	Journal of International Affairs, (Quarterly);
- Occasional Papers:	Bangladesh Journal of Law, (Bi-annual)
- Others:	Bangladesh Supreme Court Reports, (Quarterly).

Seminars and Conferences:

- Frequency	3 months
- Participation	40

- Title of a Recent Event: "Rape, as a War Crime."

Library Collections:

- Books:	4500
- Periodicals:	2480
- Newspapers:	5

Year of Establishment:	1972
Status:	Non-Governmental, Autonomous Organisation

● Bangladesh Unnayan Parishad (BUP)

33, Road No. 4, Dhanmondi R.A.
P.O.Box. 5007 (New Market), Dhaka 1205, Bangladesh.
Tel: 880-2-508097 Fax: 867021 E-mail : qka.bup@driktap.tool.nl

Chief Executive:	Q.K. Ahmad, Chairman
Other Key Personnel:	S.M. Hasanuzzaman M. Maniruzzaman Miah M.S. Alam Mia A.R. Choudhury
Number of Professional Staff:	15M, 10F
Areas of Research:	Socio Economic Issues of Development; Environment; Public Opinion Research.
Main Work at Hand:	Greenhouse Effect & Climate Change; Water Resource Development & Development - National and Regional Perspective; Public Opinion Research.

Publications:

- Journal:	*Asia Pacific Journal on Environment & Development* (Bi-annual).

	- Occasional Papers:	Public Opinion Research Reports, Research Reports.
	- Other:	A Monthly and a Quarterly Journal in Bengali.

- Recent/Forthcoming Titles:
Bangladesh: Past Two Decades & the Current Decade - Assimilating Past Experiences Towards Shaping the Future (BUP 1994); *Resources, Environment & Development in Bangladesh With Particular Reference to the Ganges, Brahmaputra & Meghna Basins* (BUP 1994); *Development & Democracy: Must there be Winners & Losers?* (BUP 1995).

Seminars & Conferences:	- Frequency:	6-8 times a year.
	- Participation:	60-100
	- Title of a Recent Event:	
	"Social Weather Survey Results"	
Library Collections:	- Books:	6500
	- Periodicals:	14
	- Newspapers:	3
	- Others:	Research Reports-300
Year of Establishment:	1980	
Status:	Private, Non-profit Organization.	

● Centre for Analysis & Choice (CAC)

House No. 65, Road No. 6A
Dhanmondi R/A, Dhaka 1209, Bangladesh.
Tel: 880-2-319919

Areas of Research:	Development of Stable Constitutional System of Government in Bangladesh; Development of Democratic Accountability Concept at all Levels; Development of Gender Balanced Politics; Enhancement of Women in Politics.
Main Work at Hand:	Legislative Support Services for Members of Parliament.

● Centre for Development Research, Bangladesh (CDRB)

55, Dhanmondi R/A, Road No. 8/A
GPO Box 4070, Dhaka 1209, Bangladesh.
Tel: 880-2-817277 Fax: 811877 E-mail : cdrb@agni.com

Chief Executive: Mizanur Rahman Shelley, Chairman

Other Key Personnel: Mohammad Enayet Karim, Director

Number of Professional Staff: 7M, 3F

Areas of Research: Socio Economic Development; Rural Development; Population, Family Planning & Human Resource Development.

Main Work at Hand: Retraining Programme for Affected Workers of Jute Mills; Support to Health Policy Formulation; Monitoring & Evaluation of Non-Formal Primary Education.

Publications: - Journal: *The Asian Affairs* (Quarterly)
- Recent/Forthcoming Titles:
The Chittagong Hill Tracts of Bangladesh: The Untold Story; Foreign Policy in a Changing World: Bangladesh, Continuity & Change (Forthcoming).

Seminars & Conferences: - Frequency: Monthly / Bi-monthly.
- Participation: 50

Scholarship, Fellowship, Training etc.:
- Title of Programme: AIBS-CDRB
- Subjects: Interdisciplinary
- Number: Varies
- Duration: 3 months to 1 Year
- Eligible: Nationals of USA & Bangladesh
- Funded by: Government of Bangladesh.

Library Collections: - Books: 2000
- Periodicals: 700
- Newspapers: 10

Year of Establishment: 1982

Status: Non-Governmental Organisation

● Centre for Policy Dialogue (CPD)
6-A Eskaton Garden, Ramya
Dhaka 1000, Bangladesh.
Tel: 880-2-837055 Fax: 835701 E-mail : rs.cpd@drik.bgd.toolnet.org

Chief Executive: Rehman Sobhan, Chairman

Other Key Personnel: Nurul Huq

Number of Professional Staff: 6M, 2F

Areas of Research:	Governance; Trade & Industries; General Economy (Sectoral Reviews).
Main Work at Hand:	Independent Review of Bangladesh's Development 1996; Governance & Development; Gender Issues & Export Sector.
Publications:	- Books; Occasional Papers: Reports on Policy Dialogue. - Recent/Forthcoming Titles: *Experiences with Economic Reform: A Review of Bangladesh's Development* 1995.

Seminars & Conferences:
- Frequency: Fortnightly.
- Participation: 20
- Title of a Recent Event:
"South Asian Dialogue."

Library Collections:
- Books   1000
- Periodicals: 300
- Newspapers: 8

Year of Establishment: 1994

Status: Non-Governmental Organisation

● Centre for Urban Studies (CUS)
Room No. 205, Science Annexe Building
Dhaka University, Dhaka 1000, Bangladesh.

Chief Executive: Nazrul Islam, Director.

Number of Professional Staff: 3M, 3F

Areas of Research:	Urbanization; Urban Poverty; Urban Environment.
Main Work at Hand:	Urban Poverty in Bangladesh; Urban Governance in Bangladesh.
Publications:	- Books; Occasional Papers - Others: Urban Studies Bulletin - Recent/Forthcoming Titles: *The Urban Poor in Bangladesh; Urban Governance in Bangladesh & Pakistan.*

Seminars & Conferences:
- Frequency: Monthly
- Participation: 20
- Title of a Recent Event:
"Urban Governance in Bangladesh & Pakistan".

Library Collections:	- Books:	500
	- Periodicals:	30
	- Newspapers:	50
	- Others	100

Year of Establishment:	1972
Status:	University Research Centre.

● **Institute of Development, Environment & Strategic Studies (IDESS)**
North South University, 12 Kemal Ataturk Avenue
Banani, Dhaka 1213, Bangladesh.
Tel: 880-2-885369 Fax: 883030

Chief Executive:	Shaukat Hassan, Director
Other Key Personnel:	Dipasis Bhadra, Director, Development.
Number of Professional Staff:	7M, 4F
Areas of Research:	Strategic Concerns; Environmental Issues; Developmental Issues.
Main Work at Hand:	Humanitarian Ceasefires Project; Sustainable Development, Environmental Security & Disarmament Interface in South Asia; Poverty Environment Linkages.

Seminars & Conferences:	- Frequency:	Monthly
	- Participation:	50
	- Title of a Recent Event:	"Training Programme for Primary Level Teachers on Environment."

Library Collections:	- Books:	6700
	- Periodicals:	66
	- Newspapers:	30

Year of Establishment:	1993
Status:	University Research Centre

● Centre for International Relations
Queen's University, Kingston, Ontario
Canada K7L 3N6.
Tel: 1-613-545-2381 Fax: 613-545-6885
E-mail : qcir@qsilver.queensu.ca

Chief Executive:	S.N. Mac Farlane, Director

Number of Professional Staff: 21M

Areas of Research:	European & Transatlantic Security; Peacekeeping; International Relations of the Former Soviet Union.
Main Work at Hand:	Multilateralism & Regional Security; Ethnic Conflict & European Security; Will NATO go East.
Publications:	- Occasional Papers: Martello Papers (6 /year) - Recent/Forthcoming Titles: *Regional Security in South Asia*
Seminars & Conferences:	- Frequency: 3/Year. - Participation: Varies. - Title of a Recent Event: "NATO Expansion".

Scholarship, Fellowship, Training etc.:

- Title of Programme:
Security & Defence Forum
- Subjects: Security.
- Number: Varies.
- Duration: 12 Months.
- Eligible: Nationals of Canada.
- Funded by: Govt. of Canada.

Year of Establishment:	1975
Status:	Non-Governmental Organisation

● Canadian Institute of Strategic Studies (CISS)
76, St. Clair Avenue West, Suit 502
Toronto, Ontario, M4V 1N2 Canada.
Tel: 1-416-964-6632 Fax: 416-964-5833
E-mail : ciss@inforamp.net

Chief Executive:	Alex Morrison, Executive Director
Other Key Personnel:	J.J. Blais, Chairman. Don MacNamara, President. James Hanson, Associate Director

Areas of Research:	Foreign and Defence Policy; Economic Strategy; Peacekeeping & Conflict Resolution.
Publications:	- Journal: *CISS Bulletin* (Quarterly).
	- Occasional Papers: Strategic Data Links; Books
	- Recent/Forthcoming Titles:
	The Canadian Strategic Forecast - Proceedings of the Annual Autumn Seminar; *Canadian Defence & Foreign Policy* - Proceedings of Annual Spring Seminar; Book on *Canadian Naval History.*
Seminars & Conferences:	- Frequency: 2/Year.
	- Participation: 100-150
	- Title of a Recent Event:
	The Canadian Strategic Forecast 1996.

Scholarship, Fellowship, Training etc.:

	- Title of Programme:
	The George G. Bell Fellowship
	- Subjects: Varies.
	- Number: 1
	- Duration: One year.
	- Eligible: Nationals of Canada
	- Funded By: CISS
Library Collections:	- Books: 1000
	- Periodicals: 300
Year of Establishment:	1976
Status:	Non-Profitable, Non-Governmental Organisation.

■ *Allahabad (UP)*

● **Society for Peace Security & Development Studies**
58 Balrampur House,
Allahabad
211001, India.
Tel: 91-532-649558 Fax: 640211

Chief Executive:	D.D. Khanna.
Number of Professional Staff:	3M

Areas of Research: South Asian Strategic Problems; Disarmament & Arms Control; Conflict Resolution; India-Pakistan Relations.

Main Work at Hand: Dialogue of the Deaf: India Pakistan Divide; Defence Versus Development.

Publications:
- Journal: *Journal of Peace, Security & Development* (Annual).
- Occasional Papers: 3/Year; Newsletters 3/Year.
- Recent/Forthcoming Titles:
Sustainable Development, Environmental Security & Disarmament Interface in South Asia.

Seminars & Conferences:
- Frequency: 2/Year
- Participation: 20
- Title of a Recent Event:
Regional Workshop on Sustainable Development.

Library Collections:
- Books: 2000
- Periodicals: 15
- Newspapers: 5

Year of Establishment: 1992

Status: Non-Governmental Organisation

■ *Calcutta (WB)*

● **Maulana Abul Kalam Institute of Asian Studies**
567, Diamond Harbour Road, Calcutta 700034
India.
Tel: 91-33-4681396

Chief Executive:	Barun De, Director.
Number of Professional Staff:	6M, 4F

Areas of Research: South Asia; Central Asia; South West Asia.

Main Work at Hand:	Nation Making in Central Asia - Uzbekistan, Tajikistan in particular; Migration, Ethnicity, Governance - in South Asia; State & Culture in South Asia - India & Bangladesh in particular.
Publications:	- Newsletter; Occasional Papers; Books. - Recent/Forthcoming Titles: *Religion, Identity & Recent History of Bangladesh; Whose Asia is it Anyway? On Asian Nationalism; Politics & Development in South Asia: India & Bangladesh.*

Seminars & Conferences:
- Frequency: 2/Year
- Participation: 20-25
- Title of a Recent Event:
"Transborder Migration from Bangladesh to West Bengal".

Scholarship,Fellowship,Training etc.:
- Title of Programme: Language Programe; Utility Fellowship.
- Subjects: Language Programe: Russian, Persian; Area Studies.
- Duration: 1 Month
- Eligible: No Restriction.
- Funded By: The Institute.

Library Collections:
- Books: 1500
- Periodicals: 120
- Newspapers: 7

Year of Establishment: 1993

Status: Government-funded; Autonomous Body.

● Netaji Institute for Asian Studies

1, Woodburn Park, Calcutta 700020
India.

Chief Executive: R. Chakraborti, Director.

Number of Professional Staff: 23M

Areas of Research: South Asia; South East Asia; Far East.

Main Work at Hand: Religion in Socio-Economic Life of India; Theories of International Relations; Bhutan: Political Economy of Development.

Publications:	- Journal:	*Asian Studies,* (Bi-annual);
	- Books	
	- Recent/Forthcoming Titles:	
	Ethno Nationalism: Indian Experience; South- *South Cooperation; The World of Thai Women.*	

Seminars & Conferences:	- Frequency:	6/7 per Year.
	- Participation:	30 - 45
	- Title of a Recent Event:	
	INA & Transfer of Power	

Library Collections:	- Books:	6000
	- Periodicals:	30
	- Newspapers:	4
	- Others	2/3 Weekly Magazines

Year of Establishment: 1981

Status: Governmental Organisation.

● School of International Relations & Strategic Studies (SIRSS)
The Director, SIRSS, P.G. Arts Building
Jadavpur University, Calcutta 700032, India.

Chief Executive Arun Kr. Banerji, Director.

Number of Professional Staff: 1M, 1F

Areas of Research: Strategic & Security Issues; Theories of International Relations Including World Order Studies; Conflict Resolution.

Main Work at Hand: Conflicts & Conflict Resolution with Particular Reference to South Asia; Issues Concerning Proliferation of Nuclear Weapons.

Publications:	- Occasional Papers; Books
	- Recent/Forthcoming Titles:
	Integration, Disintegration & World Order; *Gulf War & Energy Crisis in India.*

Seminars & Conferences:	- Frequency:	Monthly/Bi-monthly, Annually.
	- Participation:	30-35
	- Title of a Recent Event:	
	Nuclear Non-Proliferation: India's Policy.	

| Library Collections: | - Books | 350 |

Year of Establishment:	1988
Status:	Non-Governmental Organisation

● **South Asia Research Society (SARS)**
309, Jodhpur Park, Calcutta 700068
India.
Tel: 91-33-4733187 Fax: 91-33-4739175

Chief Executive:	Jayanta Kumar Ray, Director.
Other Key Personnel:	D.K. Sen Gupta, Gen. Secretary.
	Amalendu De, Vice President.
	Arun Bandyopadhyay, Joint Secretary.

Number of Professional Staff: 15M, 5F

Areas of Research:	Cooperation Broadening Measures in South Asia; Theory & Practice of Rural Development; Translation in the Field of Bengal Studies.
Main Work at Hand:	India & Nepal: Institution Building For Optimal Water Resource Management; Replication by SARS of the Bangladesh Grameen Bank Financial System in 3 Districts of West Bengal: Project Nirdhan; International Centre for Bengal Studies (ICBS) Project.
Publications:	Monthly Reports on the Rural Poverty Alleviation project.

- Recent/Forthcoming Titles:
Bengal: A Historiographical Quest; Ganatantra O Jatiyatar Agnipariksha: Bangladesh 1947-1971 (Bengali Translation of an English Book); *Bengali Language & Literatutre*, Vol.1, (Allied Publishers, Calcutta, 1996).

Seminars & Conferences:	- Frequency:	Annual
	- Participation:	30
Library Collections:	- Books	500
	- Periodicals:	9
	- Newspapers:	12
Year of Establishment:	1992	
Status:	Non-Governmental Organisation	

■ *Hyderabad*

● **American Studies Research Centre (ASRC)**
Osmania University Campus, Hyderabad 500007
India.
Tel: 91-40-7018276 Fax: 91-40-7017114
E-mail : burg@asrchyd.ernet.in

Chief Executive:	B.R. Burg, Director.
Other Key Personnel:	Sreenidhi Iyengar, Librarian.

Number of Professional Staff: 12M

Areas of Research: Any Aspect of American Studies or International Relations.

Main Work at Hand: American Civilization Course; Full Programme of International Relations Seminars, Workshops, Lectures, Continuous Programme of American Studies.

Publications: - Journal: *Indian Journal of American Studies;*
- Quarterly Newsletter.

Seminars & Conferences: - Participation: 25-35
- Title of a Recent Event:
"Economic Liberalization Indian and American Perspectives"

Scholarship, Fellowship, Training etc.:
Various Programmes
- Subjects: American Studies & International Relations
- Number: 5
- Duration:: 1 week to 4 months.
- Eligible: Indian Nationals
- Funded by: ASRC

Library Collections: - Books 12000
- Periodicals: 2026

Year of Establishment: 1964

● **Centre for Area Studies**
Osmania University, Hyderabad 500007
India.

Chief Executive: Rama S. Melkote, Director

Number of Professional Staff: 12M, 4F

Areas of Research:	Indian Ocean: Geo-Politics, Urbanization & Regional Planning; Regional Trade.
Main Work at Hand:	Population Politics of Third World Countries; SAARC-ASEAN Regional Cooperation; Peace & Security in Indian Ocean.
Publications:	- Journal; Newsletter (Bi-annual); Books

Seminars & Conferences:
- Frequency: Annual.
- Participation: 15-25
- Title of a Recent Event:
State, Globalization & Environment.

Library Collections:
- Books: 2500
- Periodicals: 10
- Newspapers: 8

Year of Establishment:	1983
Status:	University Research Centre

■ Madras (TN)

● Centre for Socio-Legal Reasearch & Documentation Service (CSRDS)
125, 5th Street, Padmanabha Nagar
Adyar, Madras, India.
Tel: 91-44-4915941 Fax: 4918195
E-mail : csrds@lwmaa.nandanet.com

Chief Executive: K. Rajakumar, Secretary.

Number of Professional Staff: 1M, 3F

Areas of Research:	Law & Society; Human Rights; Law & Weaker Section.
Main Work at Hand:	Panchayat Raj System in Practice; People Living with HIV/AIDS & Law; Children & Law.
Publications:	- Journal: *Socio-Legal Concern* (Quarterly);
	- Other publications in Tamil.

Seminars & Conferences:
- Frequency: Once in 3 Months
- Participation: NGOs, 30-35
- Title of a Recent Event:
Workshop on Tamil Nadu Panchayat Raj Act.

Library Collections:	- Books:	6000
	- Periodicals:	45
	- Newspapers:	5

Year of Establishment: 1984

Status: Non-Governmental Organisation.

● Centre for South & Southeast Asian Studies
University of Madras, Chepauk
Madras 600005, India.
Tel: 91-44-568778 Fax: 566693

Chief Executive: V. Suryanarayan, Director

Number of Professional Staff: 6M, 2F

Areas of Research: Interdisciplinary Research Programmes Relating to Sri Lanka, Maldives, Malaysia, Singapore & Indonesia; Analytical Studies Relating to India's Relations with South & Southeast Asian Countries.

Main Work at Hand: Maldives: Challenges of Development; Readings for the Study of Sri Lanka & Southeast Asia; Religion, Ethnicity & Nation Building in Indonesia.

Publications:
- Books
- Recent/Forthcoming Titles:
Kachchaitivu & Problems of Indian Fisherman in the Palk Bay Region; The Indian Ocean & it's Islands: Strategic, Scientific & Historical Perspective; South & Southeast Asia in the 1990s - Indian & American Perspectives.

Seminars & Conferences:
- Frequency: Occasional
- Participation: 50
- Title of a Recent Event:
Problems of Fisherman in the Palk Staits.

Scholarship, Fellowship, Training etc.:

	- Title of Programme:	M.Phil./Ph.D.
	- Subjects:	South & Southeast Asian Studies
	- Duration:	M.Phil: 1year; Ph.D: 3years
	- Eligible:	Nationals of India
	- Funded by:	University Grants Commission

Library Collections:	- Books	2000
	- Periodicals:	1000
	- Newspapers:	7
	- Others:	70 Dissertations.

Year of Establishment: 1977

Status: Quasi Government.

■ Mumbai (Maharashtra)

● Centre for African Studies (CAS)

Centre for African Studies, University of Mumbai
Vidyanagari (Kalina) Campus, Santacruz (E), Mumbai, India.

Other Key Personnel: Ankush B. Sawant, Director.

Number of Professional Staff: 2M, 4F

Areas of Research: Political Institution & Process; Economic Development; Social Change.

Main Work at Hand: Political Pluralism; Structural Adjustment Programme; The State in Global Context.

Publications:
- Journal: *African Currents* (Bi-annual);
- Occasional Papers; Books
- Recent/Forthcoming Titles: *India & South Africa: A Fresh Start; Area Studies in Indian Universities; Conflict Situation in Africa.*

Seminars & Conferences:
- Frequency: Fortnightly/Monthly.
- Title of a Recent Event:
Global Transformation & the Third World.

Library Collections:	- Books:	12000
	- Periodicals:	50
	- Others:	840 Microfilms

Year of Establishment: 1971

Status: Semi-Governmental, Autonomous Organisation.

● International Centre for Peace Initiatives (ICPI)

B-704, Montana, Lokhandwala Complex
Andheri West, Mumbai 400053, India.

Chief Executive: Karan R. Sawhny, Director.

Number of Professional Staff: 2M, 1F

Areas of Research:	Kashmir; India-Pakistan Relations; Conflict Resolution Process.
Main Work at Hand:	Kashmir Humanitarian Initiatives; Political Capacity Building for Peace-Keeping in South Asia; Vision 2022.

Publications: - Journal: *Peace Initiatives* (Bi-monthly).
- Occasional Papers: Track-Two Diplomacy in South Asia (Annual);
- Books
- Recent/Forthcoming Titles:
A Handbook for Conflict Resolution in South Asia; The New World Order.

Seminars & Conferences: - Frequency: varied
- Participation: 30
- Title of a Recent Event:
Resolving the Kashmir Conflict

Scholarship, Fellowship, Training etc.:
- Title of Programme: Internship
- Subjects: Regional Peace/Security.
- Number: Varies
- Duration: 2 Months.
- Eligible: Nationals of India.
- Funded By: ICPI

Library Collections: - Books: 200
- Periodicals: 25
- Others: Documents

Year of Establishment: 1990

Status: Non-Governmental Organisation

■ New Delhi

● Committee for Information & Initiative on Punjab

66, Babar Road, New Delhi 110001
India.

Key Personnel: Tapan Bose
Ram Naayan Kumar
Nitya Ramakrishnan
Ashok Agrwaal

Areas of Research:	Human Rights in Punjab: Related Legal & Political Issues.
Main Work at Hand:	State Terrorism in Punjab - A Report; Disappearences - A Documentary.
Publications:	- Recent/Forthcoming Titles: A Comprehensive Report, with Case Studies of the Human Rights Situation in Punjab with Historical Background Covering Legal, Political & Judicial Approaches Over the Last 40 Years.
Year of Establishment:	1988
Status:	Non-Governmental Organisation

● Institute for Defence Studies & Analyses (IDSA)

Sapru House, Barakhamba Road
New Delhi 110001, India.
Tel: 91-11-3317189 Fax: 3324951
E-mail : postmast@idsa.delnet.ernet.in

Chief Executive:	Jasjit Singh, Director.
Other Key Personnel:	President.
Number of Professional Staff:	22M, 15F
Areas of Research:	International Security; Disarmament Studies; Non-Proliferation.
Main Work at Hand:	CTBT: Issues & Prospects; Fissile Missile Cut-off: the Proposed Regime; Small Arms Proliferation.
Publications:	- Journal: *Strategic Analysis* (Monthly); *Strategic Digest;* - Occasional Papers; Books; News Reviews - Recent/Forthcoming Titles: *Light Weapons & International Security; Future of NPT; India-China & Panchsheel; Sea Power & Indian Security; Limited War: The Challenge of US Military Strategy.*
Seminars & Conferences:	- Frequency: 1 Every Week - Participation: 40 - Title of a Recent Event: Indo-US Workshop; India-Japan Bilateral Colloquium

216

Scholarship,Fellowship,Training etc.:

- Title of Programme:	Fellowships (3 Levels).
- Subjects:	Specific Security Related Topics.
- Number:	3
- Duration:	1-2 Years.
- Eligible:	Indian & Selected other nationals.
- Funded by:	IDSA.

Library Collections:

- Books:	37000
- Periodicals:	400
- Newspapers:	40
- Others	400 Monographs etc.

Year of Establishment: 1965

Status: Autonomous - Non-Governmental Organisation.

● International Institute for Non-Aligned Studies (IINS)

A2/59, Saddarjung Enclave
New Delhi 110029, India.
Tel: 91-11-602520   Fax: 91-11-6876294

Chief Executive: Govind Narayan Srivastava, Director General

Areas of Research: Non Aligned Movement; Human Rights, South Asian Cooperation.

Main Work at Hand: Human Rights.

Seminars & Conferences:

- Frequency:	Bi-monthly
- Participation:	100-200

Year of Establishment: 1980

Status: Non-Governmental Organisation.

■ Srinagar (J & K)

● Centre for Central Asian Studies

University of Kashmir, Srinagar 190006, India.

Chief Executive: Abdul Majid Mattoo, Director

Areas of Research: Kashmir, Lakdakh & Central Asian Countries.

Main Work at Hand: Geography of Kazakistan; Silk Routes; History of Central Asia with Special Reference to Kashmir.

| Publications: | - Journal: | *Journal of Central Asia*; |
| | - Yearly Newsletters | |

Seminars & Conferences:	- Participation:	20
	- Title of a Recent Event:	
	"Archaeological Finding at Mansbal Lake".	

Library Collections:	- Books:	4000
	- Periodicals:	20
	- Newspapers:	3

| Year of Establishment: | 1978 |

■ *Thiruvananthapuram (Kerala)*

● Institute for the Study of Developing Areas (ISDA)
4/64-2 Continental Gardens, Kowdiar
Thiruvananthapuram 695003, Kerala, India.

Chief Executive: B. Mohanan, Director.

Other Key Personnel: K. Raman Pillai, Chairman.

Number of Professional Staff: 8M, 1F

Areas of Research: Ethnic Studies; Women's Studies; Development Studies.

Main Work at Hand: Communalism & Politics in Kerala; Reservation, Policy & Social Mobility; Ecological Aspects of Housing in Kerala.

Publications: *ISDA Journal* (Quarterly); Books
- Recent/Forthcoming Titles:
Post Cold War Perspectives on International Relations; Mahatma Ghandhi's Legacy & New World Order; India's Foreign Policy in the 90s.

Seminars & Conferences:	- Frequency:	Once in 6 Months.
	- Participation:	40-50
	- Title of a Recent Event:	
	National Seminar on Re-Discovery of Ghandhi	

Scholarship, Fellowship, Training etc.:

	- Title of Programme:	
	Short Term Course in Research Methodology.	
	- Duration:	1 Month.
	- Eligible:	Nationals of India.
	- Funded By:	ISDA & ICSSR.

218

Library Collections:	- Books:	15000
	- Periodicals:	50
	- Newspapers:	12
Year of Establishment:	1991	
Status:	Non-Governmental Organisation	

JAPAN

● International Development Centre of Japan (IDCJ)

Kyofuku Building, 9-11, Tomioka 2-chome
Koto-ku, Tokyo 135, Japan.
Tel: 81-3-3630-6911 Fax: 81-3-3630-8720

Chief Executive: Saburo Kawai, Chairperson.

Other Key Personnel: Yasunobu Kawato, Executive Director
Normichi Toyomane, Director
Masaoki Takeuchi, Director
Yutaka Kurihara, Director

Number of Professional Staff: 31M, 24F

Areas of Research: Developing Economies;
Official Development Assistance;
Development Policy & Planning.

Main Work at Hand: International Joint Study on Schemes of Regional Economic Cooperation Aimed at Fostering Economic Growth in South Asia: The Role of Japan; Pakistan: Economic Development & International Assistance; Master Plan Study for Development of Southern Area of Sri Lanka.

Publications:

- Journal: *IDCJ Forum.*

- Occasional Papers: IDCJ Working Paper; IDCJ Staff Occasional Notes; IDCJ Study Summary Series.

- Others: IDCJ Quarterly News, 4/year.

- Recent/Forthcoming Titles:
The Quest for Effectiveness: A Changing Southern Africa & Japanese Economic Cooperation; Features of Japaneses Policies.

Seminars & Conferences:
- Frequency: Once a year.
- Participation: 1-13
- Title of a Recent Event:
The World Bank-Japan Research Fair.

Scholarship, Fellowship, Training etc.:

- Subjects: Development Economics; Development Policy, Project Appraisal etc.
- Number: 10-70
- Duration: 2 Days- 1 Year
- Funded By MITI ; JICA.

Library Collections:
- Books: 22000
- Periodicals: 320
- Newspapers: 8

Year of Establishment: 1971

Status: Non-Profit Institution.

● **Institute of International Relations (IIR), Sophia University**
7-1 Kioi-cho, Chiyoda-ku
Tokyo 102, Japan.
Tel: 81-3-3238-3561   Fax: 3238-3592

Chief Executive: Masatsugu Naya, Director

Number of Professional Staff: 9M, 4F

Areas of Research: International Peace & Order; Economic & Political Development in LDCs.

Main Work at Hand: Japan & Global Governance; Study on Internally Displaced Persons; Comparative Study on LDC Economies.

Publications:
- Journal: *The Journal of International Studies*, (Bi-annual).
- Occasional Papers; Research Papers Series (Irregular).
- Recent/Forthcoming Titles:
New International Studies (In Japanese); Japan & Global Governance (In Japanese).

Library Collections:	- Books:	12000
	- Periodicals:	260
	- Others	140 Occasional Papers

Year of Establishment: 1969

Status: University Research Centre.

NEPAL

■ *Kathmandu*

● Centre for Economic & Technical Studies (CETS)

P.O. Box 3174, Kathmandu
Nepal.

Chief Executive: Hari Bans Jha

Number of Professional Staff: 5M

Areas of Research: Socio Economic Studies in South Asian Perspective.

Main Work at Hand: Training Evaluation of NGOs/INGOs in Family Planning; A Study of Informal Sector Workers in Nepal.

Publications: - Recent/Forthcoming Titles:
Sustainable Development of Small Hydropower in Nepal; Nepal-India Economic Relations in the 21st Century; Empowerment of Terai Women in Nepal.

Library Collections:	- Books	2000
	- Periodicals:	10
	- Newspapers:	9

Year of Establishment: 1989

Status: Non-Governmental Organisation.

● **Centre for Economic Development & Administration (CEDA)**
P.o. Box 797, Kirtipur
Kathmandu, Nepal.
Tel: 977-1-213325   Fax: 977-1-226820

Chief Executive: Pushkar Bajracharya, Executive Director.

Number of Professional Staff: 39M, 11F

Areas of Research: Economic, Development and Environmental
                                Issues.

Main Work at Hand: Measurement of Fertility & Mortality; Trade &
                                Investment Relationaship Between Nepal &
                                Japan; Population & Environment.

Publications: - Journal: *Development Administra-*
                                                  tion (Annual)
                                - Newsletter (CEDA Samachar) & Reports

Seminars & Conferences: - Frequency: Monthly
                                - Participation: 30-100
                                - Title of a Recent Event:
                                Macro Economic Situation in Nepal.

Scholarship, Fellowship, Training etc.:

                                - Title of Programme: Macro Economic; Envi-
                                                  ronmental Management;
                                                  Rural Development.
                                - Number: 25-30
                                - Duration: 2-5 weeks
                                - Eligible: Open to all
                                - Funded by: Participants;
                                                  Central Bank of Nepal

Library Collections: - Books   18100
                                - Periodicals: 174
                                - Newspapers: 17

Year of Establishment: 1969

Status: Autonomous.

● **Centre for Nepal & Asian Studies (CNAS)**
Tribhuvan University, Kirtipur
Kathmandu, Nepal.
Tel: 977-1-231740 Fax: 227184

Chief Executive: Prem Kumar Khatry, Executive Director.

Number of Professional Staff: 22M, 5F

Areas of Research:	Political Studies; Historical Studies; Linguistics.
Main Work at Hand:	Japanese Studies; Negotiating with India.
Publications:	- Occasional Papers: Democracy Watch (Bi-annual); CNAS News-letter; Books
Library Collections:	- Books: 7000
	- Periodicals: 80
	- Newspapers: 20
	- Others: 4 News Magazines.
Year of Establishment:	1972
Status:	Non-Governmental Organisation.

● **Centre for Strategic Studies (CSIS), Sanepa, Patan.**
P.O.Box No: 8339, Kathmandu, Nepal.
Tel: 977-1-523592 Fax: 526943

Chief Executive:	Chairman
Number of Professional Staff:	7M, 3F
Areas of Research:	Global and Regional Security
Main Work at Hand:	Nuclear Proliferation and Regional Security; Proliferation and Radiological Energy
Publications:	Journal of Strategic Studies
Seminars and Conferences:	- Frequency Bi-annually
	- Participation 30-40
	- Title of Recent Event:
	"Prospect of Peace in the Middle East."

Scholarship, Fellowship, Training etc.:

	- Title of Programme:
	Environment and Security
	- Number: 1
	- Duration: 12months
	- Funded by: HIS, Netherlands
Library Collections:	- Books: 1500
	- Periodicals: 12
	- Newspapers: 15
Year of Establishment:	1983
Status:	Governmental Organisation

- **Development Associates Nepal (DEAN)**
P.O.Box 6058, Kathmandu
Nepal.
Tel: 977-1-535616

Chief Executive:	Govind D. Shrestha, Executive Director.
Other Key Personnel:	Ganesh B. Thapa, Chairman.

Number of Professional Staff: 2M

Areas of Research:	Policy; Management; Training.
Main Work at Hand:	The Role of Private Sector in Financing, Construction & Operation of Infrastructure Projects (Water Supply); Assessment of the Performance of Commercial Banks in Rural Areas.
Publications:	- Occasional Papers: People's Verdict, An Analysis of Election Results, 1994.
Seminars & Conferences:	- Frequency: 1-2/year - Participation: 50 - Title of a Recent Event: Integrated Development of Mahakali River.
Year of Establishment:	1993
Status:	Non-Governmental Organisation.

- **Himal South Asia**
G.P.O.Box 7251, Kathmandu
Nepal.
Tel: 977-1-527629   Fax: 521013
E-mail: himal@himpc.mos.com.np

Chief Executive:	Kanak Mani Dixit
Areas of Research:	Happenings in South Asia.
Main Work at Hand:	*Himal South Asia* (Monthly).
Library Collections:	- Books: 500 - Periodicals: 500
Year of Establishment:	1987
Status:	Non-Governmental Organisation.

● Institute for Integrated Development Studies (IIDS)

P.O. Box 2254, Kathmandu
Nepal.
Tel: 977-1-474718   Fax: 977-1-470831

Chief Executive:	Mohan Man Sanju, Executive Chairman.
Number of Professional Staff:	12M, 6F
Areas of Research:	Economic Study & Policy Analysis; Population & Environment; Women & Development.
Main Work at Hand:	Poverty Alleviation, Gender Equity & People's Participation:; Impact of Economic Liberalization in Nepal.
Publications:	- Journal: *Samabad* (Bi-annual).

- Occasional Papers: Policy Study Series.
- Recent/Forthcoming Titles:
The Statistical Profile of Nepalese Women: An Update in the Policy Context; Converting Water into Wealth: Regional Cooperation in Harnessing the Eastern Himalayan River; The Second Parliamentry Election.

Seminars & Conferences: - Title of a Recent Event:
Absorptive Capacity of Foreign Aid in Nepal.

Library Collections:	- Books:	1078
	- Periodicals:	25
	- Newspapers:	4
	- Others:	2153 Reports.

Year of Establishment: 1990

Status: Non-Governmental and Non-Profit Organisation.

● Nepal Centre for Contemporary Studies (NCCS)

Kupandol, Lalitpur
G.P.O. Box 3316, Kathmandu, Nepal.
Tel: 977-1-527629   Fax: 419385

Chief Executive:	Lok Raj Baral, Executive Chairman.
Number of Professional Staff:	7M
Areas of Research:	Research on Contemporary National & International Issues.
Main Work at Hand:	Assessment of Democratic Experiment in Nepal 1990-1995; Constitutional Norms and Party Behaviour in Nepal.

225

Year of Establishment: 1995

Status: Non-Governmental Organisation

● Nepal Council of World Affairs (NCWA)

NCWA Building, Harihar Bhavan, Pulchowk
Lalitpur, Kathmandu P.O. Box 2588, Nepal.

Areas of Research:	International Understanding & Cooperation Among the Nations of the World; Issues of National & Regional Concern.
Publications:	- Journal (Annual)

Seminars & Conferences:	- Frequency:	2 Talk Programmes in a Month.
	- Participation:	150
Library Collections:	- Books:	2000
	- Periodicals:	50
	- Newspapers:	4

Year of Establishment: 1948

Status: Non-Governmental Organisation.

● Nepal Foundation for Advanced Studies (NEFAS)

G.P.O. Box 6154, Kathmandu
Nepal.
Tel: 977-1-227751   Fax: 227751

Chief Executive:	Anand P. Shrestha, Executive Director.
Number of Professional Staff:	9M, 2F
Areas of Research:	Regional Cooperation; Development & Under-development; Small State Politics.
Main Work at Hand:	Development Strategy for Nepal; Political Economy of Small States; Democracy & Social Development.
Publications:	- Occasional Papers; Books - Recent/Forthcoming Titles: *Development Studies: Self Help Organizations, NGOs & Civil Society; Social Development in Nepal; Ethnic Demography of Nepal.*

Seminars & Conferences:	- Frequency:	5/Year
	- Participation:	100-150
	- Title of a Recent Event:	GATT
Library Collections:	- Books	2000

- Periodicals:	20	
- Newspapers:	4	
- Others	3	

Year of Establishment: 1992

Status: Non-Governmental Organisation

● Nepal International Centre

P.O. Box 3810, Kathmandu
Nepal.
Tel: 977-1-271812

Areas of Research:	Foreign Policy & International Affairs; Socio-Economic Issues.
Main Work at Hand:	National Agenda for Political & Economic Advancement of Nepal.
Seminars & Conferences:	- Frequency: Monthly - Participation: 50-100
Year of Establishment:	1990
Status:	Non-Governmental Organisation.

● Nepal South Asia Centre (NESAC)

G.P.O. Box 8248, Kathmandu
Nepal.
Tel: 977-1-415164 Fax: 977-1-240684

Chief Executive:	Devendra Raj Panday, Chairman.
Other Key Personnel:	Posh Raj Pandey, Secretary General; Chaitanya Mishra, Executive Member
Areas of Research:	Development Strategy; Democracy; Good Governance.
Main Work at Hand:	Endogenizing Policy Making: Series of National Dialogues on Current National Issues; Development Strategies in Nepal.
Seminars & Conferences:	- Frequency: Once a year - Participation: 45 - Title of a Recent Event: Coalition Against Corruption: Role of Civil Society
Library Collections:	- Books: 150 - Periodicals: 3 - Newspapers: 2

Year of Establishment:	1993
Status:	Non-Governmental Organisation

● **Nepal Water Conservation Foundation**
P.O.Box 2221, Kathmandu
Nepal.
Tel: 977-1-528111 Fax: 529080

Chief Executive:	Ajaya Dixit
Number of Professional Staff:	4M, 2F
Areas of Research:	Water Resources Management Research; Informal Water Education.
Main Work at Hand:	Analytical Capacity Outside of Government; Local Water Management Strategies.

Publications:
- Journal: *Water Nepal*
- Recent/Forthcoming Titles:
Natural Disasters & Logical Resilience; Political Economy of Water in South Asia (In Nepali).

Seminars & Conferences:
- Frequency: 2/year
- Participation: 70

Scholarship, Fellowship, Training etc.:
- Title of Programme: Analytical Capacity Outside Government
- Subjects: Water Resource Management
- Number: 10
- Duration: 3 Years
- Eligible: Nationals of Nepal
- Funded by: Ford Foundation, New Delhi.

Library Collections:	- Books: 1000
Year of Establishment:	1991
Status:	Non-Governmental Organisation.

228

● **Centre for Strategic Studies. (CSS)**
Victoria University of Wellington, P.O.Box 600
Wellington, New Zealand.
Tel: 64-4-4965434   Fax: 4965437

Other Key Personnel:	Terence O'Brien, Director.
Number of Professional Staff:	2M
Publications:	Working Papers (4-6 per Annum); Books; Bulletin - 4/Year
Library Collections:	- Books: 200
	- Periodicals: 6
	- Newspapers: 1
Year of Establishment:	1994

PAKISTAN

■ *Islamabad*

● **Area Study Centre for Africa, North & South America**
Quaid-i-Azam University, Islamabad 45320
Pakistan.
Tel: 92-51-825809 Fax: 821397
E-mail : rbrais@asc-qau.sdnpk.undp.org

Chief Executive:	Rasul Bakhsh Rais, Director.
Number of Professional Staff:	8M, 3F
Areas of Research:	US Politics Towards South Asian Countries, Indian Ocean & the Middle East.
Main Work at Hand:	Pakistan's Search for Security; Pakistan's Transition to Democratic Rule.
Publications:	- Journal: *Pakistan Journal of American Studies*, (Bi-annual)
	- Books

- Recent/Forthcoming Titles:
State, Society & Democratic Change in Pakistan (Edited Volume); *War Without Winners: Afghnistan's Uncertain Transition After the Cold War.*

Seminars & Conferences:	- Frequency: Once in a Year. - Participation: 150 - Title of a Recent Event: American Studies Conference
Library Collections:	- Books: 9650 - Periodicals: 77 - Newspapers: 4
Year of Establishment:	1978
Status:	University Research Centre

● Centre for Democratic Development (CDD)
1-B, Street 38, F-8/1
Islamabad, Pakistan.
Tel: 92-51-281745

Chief Executive:	I.A. Rehman
Other Key Personnel:	Kamran Ahmad
Areas of Research:	Democratic Development.
Main Work at Hand:	Studies on Democratic Functioning.
Seminars & Conferences:	- Frequency: 1/month - Participation: 60 - Title of a Recent Event: Discussion of Political Parties on Electoral Reforms
Library Collections:	- Books: 50
Year of Establishment:	1993
Status:	Non-Governmental Organisation

● Department of International Relations.
Quaid-i-Azam University, Islamabad
Pakistan.
Tel: 92-51-829391

Chief Executive:	Ijaz Hussain, Chairman.

Number of Professional Staff: 8M, 5F

Areas of Research:	South Asian Security Environment; Ethnic Conflict; Foreign Policy of India and Pakistan.
Seminars & Conferences:	- Frequency: Fortnightly - Participation: 50-60 - Title of a Recent Event: National Seminar on UN.
Year of Establishment:	1973
Status:	University Research Centre

● Institute for Strategic Studies, Islamabad (ISSI)

P.O. Box 1173, Sector F-F/2, Islamabad
Pakistan.
Tel: 92-51-9204423 Fax: 92-51-9204658

Chief Executive:	Agha Murtaza Pooya, Chairman.
Number of Professional Staff:	6M, 5F
Areas of Research:	International Issues of Peace & Security.
Main Work at Hand:	Quarterly Workshops on South Asia, Central Asia; Afghanistan & Issues of Peace & Security; Bilateral Fora with China, US & Canada; International Conference on Peace in Asia.
Publications:	- Journal: *Strategic Studies* (Quarterly); *Strategic Perspective* (Quarterly); - Occasional Islamabad Papers.
Seminars & Conferences:	- Frequency: Varies - Participation: Varies - Title of a Recent Event: Workshop on Middle East Peace Process
Library Collections:	- Books: 12000 - Periodicals: 130 - Newspapers: 10
Year of Establishment:	1973
Status:	Autonomous.

● Institute of Policy Studies (IPS)

Nasr Chambers, Block 19, Markaz F-7
Islamabad 44000, Pakistan.
Tel: 92-51-818230 Fax: 824704
E-mail : postbox@ips.isb-erum-com.pak

Chief Executive:	Khurshid Ahmad, Chairman.
Other Key Personnel:	Khalid Rahman, Executive Director.
Number of Professional Staff:	79M, 3F
Areas of Research:	Pakistani Society & Politics; Economy; Education.
Main Work at Hand:	Pakistan-India Relations; Poverty Alleviation; Islamisation of Laws & Economics: Case Study of Pakistan.
Publications:	- Journal; Books; Newsletters - Recent/Forthcoming Titles: *Mass Resistance in Kashmir; Islamic Resurgence; New World Order.*

Seminars & Conferences:
- Frequency: Monthly
- Participation: 50-150
- Title of a Recent Event:
International Seminar on Poverty Alleviation in Pakistan.

Library Collections:
- Books: 9800
- Periodicals: 200
- Newspapers: 9

Year of Establishment:	1979
Status:	Non-Governmental Organisation.

● Institute of Regional Studies

Nafdel Complex, 56-F Blue Area, Nazimuddin Road
F-6/1, Islamabad 44000, Pakistan.
Tel: 92-51-920974 Fax: 92-51-9204055
E-mail : aziz@irspak.sdnpk.undg.org

Chief Executive:	Nishat Ahmad, President.
Other Key Personnel:	Bashir Ahmad, Senior Fellow.
Number of Professional Staff:	4M, 7F
Areas of Research:	Foreign Affairs: South/South West Asia; South Asian Politics; South Asian Security & Defence.
Main Work at Hand:	Indian Elections; The Tamil Question in Sri Lanka:Indo-Sri Lankan Relations; The Crisis of State Legtimacy in Afghanistan.
Publications:	- Journal: *Regional Studies* (Quarterly).

	- Occasional Papers:	Regional Perspective; Spotlight (Monthly); Selections from Regional Press (Fortnightly);
	- Books	
	- Recent/Forthcoming Titles:	

Terror in Indian Held Kashmir: Massive Violation of Human Rights; Uprising in Indian Held Jammu & Kashmir, 1991; Indian Political Scene, 1989.

Library Collections:	- Books:	7000
	- Periodicals:	87
	- Newspapers:	28
Year of Establishment:	1982	
Status:	Autonomous.	

● Pakistan Institute of Development Economics (PIDE)
P.O. Box 1091, Islamabad, Pakistan.

Chief Executive:	Director
Number of Professional Staff:	46M, 19F
Areas of Research:	Economics; Agriculture; Demography.
Publications:	- Journal: *Pakistan Development Review* (Quarterly)
	- Research Reports; Books
	- Recent/Forthcoming Titles:

Structural Change in Pakistan's Agriculture; On Raising the Level of Economic & Social Well-Being of the People (1992); Development Economics: A New Paradigm (1993).

Seminars & Conferences:	- Frequency:	Seminars 20/year; Annual Conference.
	- Participation:	30-400
	- Title of a Recent Event:	
	Poverty in South Asia.	
Scholarship, Fellowship, Training etc.:		
	- Subjects:	Economics & Demography.
	- Duration:	Annual.
	- Eligible:	Nationals of Pakistan
	- Funded by:	Australian University.
Library Collections:	- Books:	30000

- Periodicals:	300
- Newspapers:	12
- Others	6200

Year of Establishment: 1957

Status: Semi-Governmental Organisation.

● **Pakistan Security & Development Association (PASDA)**
P.O. Box 2306, Islamabad
Pakistan.

Chief Executive: Niaz A. Naik, Secretary General

Other Key Personnel: Wasim Sajjad (Chairman)
Ashraf W. Taban, Munir Ahmad Khan,
K.M. Arif, Mufti Abbas and Pervaiz Iqbal
Cheema

Areas of Research: Arms Control; Regional Security;
Economic & Social Development.

Year of Establishment: 1996

Status: Non-Governmental Organisation.

● **Pakistan Institute for Environment Development Action Research (PIEDAR)**
2nd Floor, Yasin Plaza, 74-W, Blue Area
Islamabad, Pakistan.
Tel: 92-51-820454 Fax: 92-51-276507

Chief Executive: Syed Ayub Qutub, President.

Number of Professional Staff: 6M, 5F

Areas of Research: Environment Management; Sustainable Development; Community Empowerment.

Main Work at Hand: Community Based Irrigation Management Project, Punjab; Environmental Management in Low Income Urban Areas; Primary Education for Rural Girls.

Publications: - Occasional Papers: 6/year; Newsletters (Bi-annual); Reports (Quarterly).
- Recent/Forthcoming Titles:
.*Agriculture Policy Analysis for Pakistan.*

Year of Establishment: 1992

Status: Non-Governmental Organisation.

● **Sustainable Development Policy Institute (SDPI)**
No. 46 Street 12, F-6/3
Islamabad, Pakistan.
Tel: 92-51-278134/36 Fax: 278135
E-mail : main@sdpi.sdnpk.undp.org

Chief Executive:	Shahrukh Rafi Khan, Director.
Number of Professional Staff:	11M, 5F
Areas of Research:	Sustainable Agriculture; Macro Economic Policy & Structural Adjusment; Health, Education & Governance.
Main Work at Hand:	Sustainable Cotton Production; Privatisation of Drinking Water Supply Schemes.
Publications:	- Recent/Forthcoming Titles: *Habitat II NGO Country Report; Anthology of Papers from the NGO Summit; Proceedings of the OFWM Conference.*

Seminars & Conferences:
- Frequency: At Least Twice a Year.
- Title of a Recent Event:
Green Economics Conference.

Library Collections:
- Books: 6556
- Periodicals: 100
- Newspapers: 8

Year of Establishment:	1992
Status:	Non-Governmental Organisation.

■ *Jamshoro*

● **Department of International Relation**
University of Sindh
Jamshoro, Pakistan.
Tel: 92-221-77168

Chief Executive:	Chachar Abdul Khalique, Chairman
Number of Professional Staff:	7M, 3F
Areas of Research:	Nuclear Proliferation; South Asian Security; Strategic Studies.

Library Collections:
- Books: 2300
- Periodicals: 6
- Newspapers: 4

Year of Establishment:	1973
Status:	Semi Governmental, Autonomous Organisation.

■ Karachi

● Department of International Realations
University of Karachi, Karachi
Pakistan.

Chief Executive:	Talat A. Wizarat, Chairperson.
Number of Professional Staff:	6M, 10F
Areas of Research:	Middle East; South Asia; Central Asia.
Main Work at Hand:	Confidence Building Measures and Conflict Resolution in South Asia & Middle East.
Publications:	- Recent/Forthcoming Titles: *Contemporary Central Asia; Pakistan's Security Concerns & Regional Peace.*
Seminars & Conferences:	- Frequency: Twice a year, one International & one National.
Library Collections:	- Books 600
Year of Establishment:	1958

● Pakistan Institute of International Affairs (PIIA)
Aiwan-e-Sadar Road, P.O. Box 1447, Karachi 74200, Pakistan.

Chief Executive:	Fatehyab Ali Khan, Chairman.
Other Key Personnel:	Syed Adil Hussain, Secretary & Editor.
Number of Professional Staff:	5M
Areas of Research:	International Issues with Special Reference to Pakistan.
Publications:	- Journal: *Pakistan Horizon* (Quarterly); - Special Issues of Pakistan Horizon, (Annually); Books - Recent/Forthcoming Titles: *UN at 50*
Seminars & Conferences:	- Frequency: Twice a month
Library Collections:	- Books: 27000 - Periodicals: 180 - Newspapers: 15
Year of Establishment:	1947

■ *Lahore*

● **Aurat Foundation**
8-B LDA, Garden View Apartments
Lawrence Road, Lahore, Pakistan

Chief Executive:	Nigar Ahmed
Number of Professional Staff:	36M, 33F
Areas of Research:	Major Concerns of Women in Pakistan - Health, Education, Legal Rights, Violence Against Women and Political Representation.
Publications:	Newsletter (Quarterly).

Library Collections:		
	- Books	2600
	- Periodicals:	300
	- Newspapers:	6
	- Others	1000 Research Report

Year of Establishment:	1986
Status:	Non-Governmental Organisation

● **Centre for South Asian Studies (CSAS)**
Quaid-i-Azam Campus, New Campus
Punjab University, Lahore, Pakistan.

Chief Executive:	Sarfarz Hussain Mirza, Director
Other Key Personnel:	Rafique Ahmed, Advisor
Number of Professional Staff:	5M, 2F
Areas of Research:	Economic, Political & Social Developments in South Asia.
Main Work at Hand:	WTO & India-Pakistan Relations; India's Election Scenario - Implications for Peace & Security in Asia; UN Resolutions & Kashmir.

Publications:		
	- Journal:	*South Asian Studies* (Bi-annual)
	- Occasional Papers; Books	
	- Recent/Forthcoming Titles: *India's Election Scenario.*	

Seminars & Conferences:		
	- Frequency:	Varies

Library Collections:		
	- Books	10000
	- Periodicals:	50
	- Newspapers:	10

Year of Establishment: 1976

Status: Semi-Government.

● Department of Political Science, Punjab University, Lahore
New Campus, Punjab University
Lahore, Pakistan.
Tel: 92-42-5863982

Chief Executive: Sajjad Naseer, Chairman

Number of Professional Staff: 10M, 2F

Areas of Research: Pakistan's Foreign Relations; Political System of Pakistan; Regional Organizations.

Seminars & Conferences:
- Frequency: Varies
- Participation: 40
- Title of a Recent Event:
Quaid-i-Azam Memorial Lecture.

Library Collections:
- Books 10444
- Periodicals: 14
- Newspapers: 5

Year of Establishment: 1933

Status: Semi-Governmental Organisation.

● Human Rights Commission of Pakistan (HRCP)
13, 3rd Floor, Sharif Complex
Main Market, Gulberg II, Lahore 54660, Pakistan.
Tel: 5750217 Fax: 5713078

Chief Executive: Asma Jahangir, Chairperson

Other Key Personnel: I.A. Rehman, Director.
Zohra Yusuf, Secretary General

Number of Professional Staff: 17M, 2F

Areas of Research: Women; Children; Minorities.

Main Work at Hand: Women's Rights; Penal System; Democratic Development.

Publications:
- Newsletter (Quarterly); *Jehd-i-Haq* (Monthly)
- Occasional Papers: Reports of Seminar, 4/year; Report of Fact Finding Mission 3/year.

- Recent/Forthcoming Titles:
Study of Elections 1993; Review of the Consti-

tution of Pakistan; Press: Freedom & Responsibility.

Seminars & Conferences:	- Frequency: 10/year
	- Participation: 50-200
	- Title of a Recent Event:
	Accountability of State Institutions.

Library Collections:
- Books: 200
- Periodicals: 40
- Newspapers: 11
- Others Reports.

Year of Establishment: 1986

Status: Non-Governmental Organisation.

■ *Peshawar*

● **Area Study Centre, Central Asia**
University of Peshawar, Peshawar
Pakistan.

Chief Executive: Azmat Hayat Khan, Director.

Number of Professional Staff: 7M

Areas of Research: Central Asian Republics; Afghanistan; Sinkiang Province of China.

Main Work at Hand: Mineral Potential of Kazakhstan; Common Market Concept in Terms of Central Asia; Road Links to Central Asia.

Publications:
- Journal: *Central Asia* (Monthly).
- Books.
- Recent/Forthcoming Titles:
Socio Economic & Political Factors of Afghanistan Migration to Pakistan; Constitutional Development in Afghanistan.

Seminars & Conferences:
- Frequency: Weekly
- Title of a Recent Event:
International Seminar on Central Asia.

Scholarship, Fellowship, Training etc.:
- Subjects: Central Asia
- Number: 8
- Duration: 9
- Eligible: Open to All
- Funded by: UGC Federal Government.

239

Library Collections:	- Books:	13059
	- Periodicals:	10
	- Newspapers:	9

Year of Establishment: 1974

Status: University Research Centre

● Department of International Relations
University of Peshawar, Peshawar
Pakistan.
Tel: 92-521-42363

Chief Executive: Adnan Sarwar Khan, Chairman.

Number of Professional Staff: 5M, 1F

Areas of Research: Afghanistan; China; South Asia.

Main Work at Hand: Politics & Foreign Policy of South Asian Countries.

| **Seminars & Conferences:** | - Frequency: | Fortnightly Seminars. |
| | - Participation: | 100 |

Library Collections:	- Books	5000
	- Periodicals:	600
	- Newspapers:	5

Year of Establishment: 1984

Status: University Research Centre

● Department of Political Science
University of Peshawar, Peshawar
Pakistan.
Tel: 92-91-42292

Chief Executive: F.A. Durrani, Chairman

Other Key Personnel: Iqbal Tajik

Number of Professional Staff: 1M, 1F

Areas of Research: Indo-Pakistan Relations; Pakistan-US Relations; SAARC.

Year of Establishment: 1962

Status: University Research Centre

■ *Rawalpindi*

● Christian Study Centre (CSC)
126-B Murree Road, P.O. Box 529
Rawalpindi, Pakistan.
Tel: 92-51-567412 Fax: 568657

Chief Executive:	Dominic J. Moghal, Director
Other Key Personnel:	Christine Amjad-Ali Jennifer Jevan Mehbab Sada

Number of Professional Staff: 5M, 4F

Areas of Research: Christian-Muslim Relations & Islam; Minorities in Pakistan; Contextual Christian Theology.

Main Work at Hand: Joint Electoral System for Minorities in Pakistan; Promoting Tolerence & Pluralism through Mass Media.

Publications:
- Journal: *Al-Mushir* (Quarterly).
- Recent/Forthcoming Titles:
Passion for Change; The Legislative History of Shariah; The Christian Minority in the North West Frontier Province of Pakistan.

Seminars & Conferences:
- Participation: 30-35
- Title of a Recent Event:
Minorities Rights & Tolerence.

Scholarship, Fellowship, Training etc.:
- Title of Programme: Islam in Pakistan
- Subjects: Pakistani Islam: Social, Cultural & Political Aspects.
- Number: 2
- Duration: 8 Weeks
- Eligible: Nationals of All Countries
- Funded by: Institution from which Students Come.

Library Collections:
- Books: 12000
- Periodicals: 124
- Newspapers: 8

Year of Establishment: 1967

Status: Non-Governmental Organisation.

● **Foundation for Research on International Environment, National Development and Security (FRIENDS)**
Friends Centre, 88, Race Course Scheme Race Course Road
Street 3, Rawalpindi Cantt, Pakistan.
Tel: 92-51- 518331 Fax: 9251-564244
E-mail : syed%friends@sdnpk.undp.org

Chief Executive:	Mirza Aslam Beg, Chairman
Other Key Personnel:	Fasahat H. Syed, Vice Chairman
Number of Professional Staff:	10M
Areas of Research:	Regional Economic Cooperation; Nuclear Disarmament and Conventional Arms Control Including Light Weapons; Fundamental Economic Right.
Main Work at Hand:	OIC and the Contemporary Issues of the Muslim World; Nuclear Disarmament and Conventional Arms Control Including Light Weapons.

Publications:
- Journal: *National Development and Security* (Quarterly);
- Books
- Recent/Forthcoming Titles:
Security, Trade and Advanced Technologies in South Asia: Opportunities and Strategies for Regional Cooperation; Culture of Peace in Central and South Asia.

Seminars & Conferences:	- Frequency:	10/year
	- Participation:	50 to 200
Library Collections:	- Books:	1139
	- Periodicals:	7
	- Newspapers:	20
Year of Establishment:	1991	
Status:	Non-Governmental Organisation.	

242

■ *Colombo*

● **Bandaranaike International Diplomatic Training Institute (BIDTI)**
Suite 3G-07, BMICH, Bauddhaloka Mawatha
Colombo 07, Sri Lanka.
Tel: 94-1-682109 Fax: 682111

Chief Executive:	Vernon L.B. Mendis, Director General.
Other Key Personnel:	Lorna S. Dewaraja, Director.
Areas of Research:	World Affairs; Professional Diplomacy - Institutions, Procedures, Techniques; Internationalism - Systems, Structures, Trends.
Main Work at Hand:	Conduct of Training Programme in Diplomacy & International Affairs; Programmes of Special Courses on Selected Fields.
Publications:	Monthly News Letter
Library Collections:	- Books: 150 - Periodicals: 15
Year of Establishment:	1995
Status:	Non-Governmental Organisation.

● **Centre for Policy Research and Analysis (CEPRA)**
Faculty of Law, University of Colombo
Colombo 03, Sri Lanka.
Tel: 94-1-595667 Fax: 7868297
E-mail : cepra@sri.lanka.net

Chief Executive:	Jayadeva Uyangoda and N. Selvakumaran, Co-Directors
Other Key Personnel:	W.D. Lakshman
Number of Professional Staff:	6M, 4F
Areas of Research:	Constitutional & Legislative Reform; Conflict Management & Resolution.
Main Work at Hand:	Oral History Transcripts; Documents on Media & Reporting on Conflict.
Publications:	- Journal: "Focus" - Thrice Annually. - Recent/Forthcoming Titles: *Swiss Federalism: Lessons for Sri Lanka;* *Essays on Constitutional Reform; Centre* *Provincial Relations in Canada - Lessons for* *Sri Lanka.*

Seminars & Conferences:	- Frequency:	10-12 Per Year.
Library Collections:	- Books:	500
	- Periodicals:	10
Year of Establishment:	1993	
Status:	Semi-Governmental Organisation.	

● Institute of Policy Studies (IPS)
No. 99 St. Michael's Road, Colombo 03
Sri Lanka.
Tel: 94-1-431368 Fax: 431395
E-Mail : ips@sri.lanka.net

Chief Executive:	Saman Kelegama, Executive Director.
Number of Professional Staff:	12M, 6F
Areas of Research:	Macro-Economics; International Economics; Agricultural Economics.
Main Work at Hand:	Scheme for Regional Economic Cooperation at Fostering Economic Growth in South Asia: Role of Japan; Financial Resources Mobilization in Health Sector in Sri Lanka; Impact of Labour Legislation on Labour Demand.
Publication:	- Recent/Forthcoming Titles: *How Open has the Sri Lankan Economy Been?; Demand for Money & Inflation in Sri Lanka; Producer Price Statistics for Staple Food Crops in Sri Lanka: The Case of Paddy.*

Seminars & Conferences:	- Frequency:	(Bi-monthly)
	- Participation:	35-40
Library Collections:	- Books:	3000
	- Periodicals:	12
	- Newspapers:	7
Year of Establishment:	1990	
Status:	Non-Governmental Organisation.	

● International Centre for Ethnic Studies (ICES)
2 Kynsey Terrace, Colombo 08
Sri Lanka.
Tel: 94-1-698048, 08-34892 Fax: 696618
E-Mail : ices_cmb@sri.lanka.net

Chief Executives:	Radhika Coomaraswamy, Director Kingsley De Silva, Director (Kandy branch)

Number of Professional Staff: 5M, 6F

Areas of Research: Ethnic Studies; Conflict Resolution; Violence Against Women.

Main Work at Hand: Comparative Federalism; Multiculturalism & Ethnic Coexistence in South & South East Asia; Devolution in Sri Lanka.

Publications:
- Journal: *The Thatched Patio* (Quarterly); *Ethnic Studies Report* (Quarterly).
- Books
- Recent/Forthcoming Titles: *Ideology & the Constitution: Essays on Constitutional Jurisprudence (1996); Sri Lanka: Devolution Debate (1996).*

Seminars & Conferences:
- Frequency: Monthly
- Participation: 30-50

Library Collections:
- Books: 6500
- Periodicals: 50
- Newspapers: 12
- Others: 25

Year of Establishment: 1982

Status: Non-Governmental Organisation.

● Marga Institute

93/10 Dutugemunu Street, P.O. Box 601
Kirillapone, Colombo 06, Sri Lanka.
Tel: 94-1-828544 Fax: 828597
E-mail : marga@sri,lanka.net

Chief Executive: Godfrey Gunatilleke, Executive Chairman.

Number of Professional Staff: 16M, 9F

Areas of Research: Economic Studies; Social Studies; International Studies.

Main Work at Hand: Equity in Health; Brain Drain from Sri Lanka.

Publications:
- Journal: *Marga* (Quarterly).
- Occasional Papers, Mimeographs, Seminar Papers; Books.

245

	- Others:	*Developement Review,* Quarterly Newsletter.
	- Recent/Forthcoming Titles:	
	Collective Identities, Nationalism & Protests in Sri Lanka.	
Seminars & Conferences:	- Frequency:	Quarterly
	- Participation	35
Library Collections:	- Books:	20000
	- Periodicals:	65
	- Newspapers:	12
	- Others:	50
Year of Establishment:	1972	
Status:	Non-Governmental Organisation.	

● Regional Centre for Strategic Studies (RCSS)

4-101, BMICH, Bauddhaloka Mawatha
Colombo 7, Sri Lanka.
Tel: 94-1-688601   Fax: 688602
E-mail: rcss@sri.lanka.net

Chief Executive:	Iftekharuzzaman, Executive Director
Other Key Personnel:	John Gooneratne, Associate Director
Number of Professional Staff:	2M, 3F
Areas of Research:	Cónceptions and problems of national and regional security in South Asia; Conflicts and conflict resolution in South Asia; Politics, development, governance and regional cooperation in South Asia.
Main Work at Hand:	Annual Winter Workshop on "Sources of Conflict in South Asia: Ethnicity, Refugees, Environment"; Annual Summer Workshop on "Defence, Technology and Cooporative Security in South Asia"; Several collaborative studies on themes of current interest.
Publications:	RCSS Policy Studies (4-6/year); Newsletter (Quarterly); Books
	- Recent Forthcoming Titles:
	Refugees and Regional Security in South Asia; Diplomacy & Domestic Politics in South Asia; Regional Economic Trends and South Asian Security; Ethnicity and Constitutional Reform in South Asia.

Seminars and Conferences:	- Frequency:	2-4/year
	- Participation:	35-70
Library Collections:	- Books:	1200
	- Periodicals:	20
Year of Establishment:	1992	
Status:	Non-Governmental, Non-Profit, Regional Organisation.	

● **Social Scientists Association**
425/15 Thimbirigasyaya Road, Colombo 05
Sri Lanka.
Tel: 94-1-501339

Chief Executive:	Kumari Jayawardena, Secretary.
Number of Professional Staff:	2M, 5F
Areas of Research:	Current Politics; Ethnicity; Gender.
Publications:	- Journal:	*Pravada* (Monthly).
Seminars & Conferences:	- Frequency:	Annual
	- Participation:	40
	- Title of a Recent Event:	
	Subaltern Conference	
Library Collections:	- Books:	2500
	- Periodicals:	4
	- Newspapers:	4
Year of Establishment:	1979	
Status:	Non-Governmental Organisation	

247

UNITED STATES OF AMERICA

● **Association of Third World Studies, Inc.**
Dept. of Economics, Weber State
University, 3807 University Circle, Ogden, USA.
Tel: 8016267442

Chief Executive:	John Mukum Mbaku, President.
Other Key Personnel:	Zia H. Hashmi, Executive Director. Harold Isaacs, Treasurer. Chaitram Singh, Secretary.
Areas of Research:	All Aspects of Third World Studies.
Main Work at Hand:	National Development, Imperialism & Religion in the Third World; The Third World on the Brink of the Twenty-First Century: Problems, Prospects & Solutions.
Publications:	- Journal: *Journal of Third World Studies*, 2/Year. - Occasional Papers; Proceedings of Annual Meetings; Newsletter. - Recent/Forthcoming Titles: *Multiparty Democracy & Political Change: Constraints to Democratization in Africa; Civil-Military Relations in the Post Cold War Era* (A Series of Volumes).
Seminars & Conferences:	- Frequency: Annual - Title of a Recent Event: National Development, Imperialism & Religion.
Year of Establishment:	1984
Status:	Non-Profit Organisation.

● **Joan B. Kroc Institute for International Studies**
University of Notre Dame, P.O. Box 639
Notre Dame, Indiana 46556, USA.
Tel: 1-219-6316970 Fax: 219-6316973 E-mail : krizmanich.l@nd.edu

Chief Executive:	Raimo Vayrynen, Director.
Number of Professional Staff:	9M, 2F
Areas of Research:	International Institution & Peacekeeping; Conflicts & Conflict Theory; Transnational Social Movements; Cooperative Security.
Main Work at Hand:	Early Warning & Prevention of Conflicts; UN Enforcement & Peacekeeping; Economic Sanctions; Transnational Human Rights Organization.

248

Publications:	- Bi-annual Report ; Annual Report. - Recent/Forthcoming Titles: *Development Ethics; India & the Bomb;* *Economic Sanctions.*
Seminars & Conferences:	- Frequency: 1/year. - Participation: 50-80
Year of Establishment:	1986
Status:	University Research Centre.

● Programme in Arms Control, Disarmamant & International Security (ACDIS)

ACDIS, University of Illinois,
Urbana - Champaign, 505 E Armory Ave Rm 35, Champaign IL 61820, USA.

Chief Executive:	Stephen P. Cohen, Director.
Other Key Personnel:	Merrily Shaw
Number of Professional Staff:	15M, 7F
Areas of Research:	US Policy in South Asia; Arms Control & Disarmamnt.
Main Work at Hand:	Analysis of the Commercial Value of Indigenous Plutonium Reprocessing & Uranium Enrichment in India & Pakisan; A Re-examination of the 1990 Crisis in South Asia; Regional Conflict Resolution Strategies.
Publications:	- Journal: *Sword & Ploughshares* (Quarterly). - Occasional Papers: ACDIS Occassional Paper Series - Others: ACDIS News (Quarterly); ACDIS Research Reports. - Books
Seminars & Conferences:	- Frequency: 2-6 Per Year. - Participation: 10-50 - Title of a Recent Event: Technology for Peace.
Scholarship, Fellowship,Training etc.:	- Title of Programme: "ACDIS Interdisciplinary Fellowship". - Number: 3

	- Duration:	2 Semester
	- Eligible:	Current Graduate Students at UIUC
	- Funded by:	University of Illinois.
Library Collections:	- Books:	2028
	- Periodicals:	211
	- Newspapers:	9
	- Others:	50
Year of Establishment:	1978	
Status:	University Research Centre.	

● Programme in South & West Asian Studies, University of Illinois

221 International Studies Building, 910, 5th Street
Champaign, IL 61820, USA.

Chief Executive:	Marvin G. Weinbaum, Director.
Number of Professional Staff:	3M, 2F
Areas of Research:	Regional Security Issues; Regional Political Economy.
Main Work at Hand:	Security Concerns South, West & Central Asia; The Interaction of Economic & Political Pluralism.
Publications:	- Recent/Forthcoming Titles: *South Asia Approaches the Millennium: Re-examining National Security* (1995).
Seminars & Conferences:	- Frequency: 1/year.
	- Participation: 30
	- Title of a Recent Event: Consequences of the Religious Conflict.
Year of Establishment:	1983

● South Asia Group for Action & Refelection (SAGAR)

6921, Creekside Road, Downers Grove
Illinois 60516-3434, USA.
Tel: 1-708-9642258

Chief Executive:	Harinder S. Lamba, Coordinator.
Other Key Personnel:	S. Naqi Akhter; Dinesh Sampat (Coordinators).
Areas of Research:	Regional Cooperation (South Asia); Human Development & Empowerment; Environmental Rejuvenation.

250

Main Work at Hand:	India-Pakistan Cooperation; Equitable Education Policy; Alternative Development Areas.
Seminars & Conferences:	- Frequency: Annual - Participation: 30-50 - Title of a Recent Event: India-Pakistan Amity.
Library Collections:	- Books: 100 - Periodicals: 8 - Newspapers: 1
Year of Establishment:	1993
Status:	Non-Governmental Organisation.

● The Asia Foundation's Centre for Asian Pacific Affairs
465, California Street, 14th Floor
San Francisco, CA 94104, USA.

Chief Executive:	William P. Fuller, President.
Other Key Personnel:	Richard Wilson, Executive Director Richard Fuller, Area Director
Areas of Research:	Security; Human Rights; Economic Regionalism.
Publications:	- Occasional Papers: 4-5/year, - Recent/Forthcoming Titles: ASEAN'S Security Framework.
Seminars & Conferences:	- Frequency: 5-6/Year - Participation: 20-50 - Title of a Recent Event: Human Rights Workshop.
Library Collections:	- Books: 300 - Periodicals: 20 - Newspapers: 5
Year of Establishment:	1941
Status:	Non-Governmental Organisation.

● The Henry L. Stimson Centre
21 Dupont Circle, NW, 5th Floor
Washington, DC 20036, USA.
Tel: 1-202-2235956 Fax: 202-7859034
E-mail : info@stimson.org

Chief Executive:	Micheal Krepon, President
Other Key Personnel:	Barry Blechman, Chairman

251

Number of Professional Staff: 7M, 9F

Areas of Research: Confidence Building; US Foreign Policy; Preventative Diplomacy.

Main Work at Hand: A Handbook of Confidence Building Measures for Regional Security; Crisis Prevention, Confidence Building, & Reconciliation in South Asia; An Evolving US Nuclear Posture, Project on Eliminating Weapons of Mass Destruction.

Publications: - Occasional Papers; Books; 2 newsletters; 20 Reports; 3 Handbooks.
- Recent/Forthcoming Titles:
UN Peace Keeping, American Policy & the Uncivil Wars of the 1990's; Regional Confidence Building in 1995: South Asia, The Middle East & Latin America; Confidence Building Measures in South Asia.

Seminars & Conferences:
- Frequency: 3/Month
- Participation: 10-50

Scholarship,Fellowship,Training etc.:
- Title of Programme: "Visiting Fellows Programme."
- Subjects: Confidence Building in South Asia.
- Duration: 2 months.
- Eligible: Nationals of India, Pakistan & China.
- Funded by: Ford Foundation.

Library Collections:
- Books: 500
- Periodicals: 30
- Newspapers: 5

Year of Establishment: 1989

Status: Non-Governmental Organisation.

● University of Wisconsin

University of Wisconsin, Oshkosh, WI 54901
USA.
Tel: 414-424-0929 Fax: 414-424-7317
E-mail : khan@vaxa.cis.uwosh.edu

Chief Executive: Zillur R. Khan, Chairperson.

Other Key Personnel: John E. Kerrigan, Chancellor.

Number of Professional Staff: 12M, 7F

Areas of Research: Intra & Inter Regional Cooperation; Constitutionalism & the Democratizing Process; Revolution & Mass Movements.

Main Work at Hand: Constitution, Caretaker Government & the Democracy; World Bank IMF & the Disadvantages; Regional Cooperation & the Changing Concepts.

Publications: - Recent/Forthcoming Titles:
SAARC & the Superpowers, The Third World Charismat.

Seminars & Conferences:
- Frequency: Once every three years
- Participation: 60-80
- Title of a Recent Event:
Bengal Studies Conference.

Scholarship, Fellowship, Training etc.:
- Title of Programme: Bilateral Faculty Exchange Programme.
- Subjects: American & Bangladesh Studies.
- Number: 2
- Duration: 3 Months
- Eligible: Americans & Bangladeshis

● **War & Society in South Asia Group (WASSAG)**
University of Illinois, 359, Armory Building
Champaign, IL 61820, USA.

Chief Executive: DeWitt Ellinwood, Stephen P. Cohen, Co-Directors.

Areas of Research: South Asian Military History & the Role of the Armed Forces in South Asian Society & Development, Both Contemporary & Historical Context.

WASSAG was founded in 1994 by a Group of American, Indian, Canadian & Pakistani Scholars: It is Open to Anyone Interested in South Asian Military History.

Year of Establishment: 1993

Status: University Research Center.